IMPROVING ON NATURE

IMPROVING ON NATURE

The Brave New World of Genetic Engineering

ROBERT COOKE

A DEMETER PRESS BOOK

Quadrangle/The New York Times Book Co.

Library of Congress Cataloging in Publication Data

Cooke, Robert, 1935–
   Improving on nature.

   Includes index.
   1. Genetic engineering. I. Title.
QH442.C66   301.24'3   76–50815
ISBN 0–8129–0667–5

Dedicated to Sue, Greg, Karen and Emily;
and to the good friends in Massachusetts and California
who make all things possible.

Contents

IMPROVING ON NATURE

Genes, Genesis, New Horizons

DARK YEARS of war and fear had set the scene as a small, secret band of scientists nervously pulled control rods from the heart of their first primitive atomic pile, thus awakening the nuclear genie, the radioactive beast so carefully hidden beneath the stands of an empty football stadium at the University of Chicago.

Unaware of the dangers, Chicago's people had no way of knowing during this first, hesitant unleashing of the atomic chain reaction that it was the quiet beginning of a whole new era. Unfortunately, it was an era which would be proclaimed loudly, in terrifying terms, with the cataclysmic destruction of Hiroshima on August 6, 1945, and Nagasaki three days later.

History, of course, is still collecting and sifting the consequences of that first controlled fissioning of the atom in the heart of Chicago. Yet even while the historians continue their scribbling, another era of equal or even greater promise and peril is dawning; and it, too, concerns more than just the folks in Chicago. Indeed, most nonscientists aren't fully aware of what's going on, and may not hear much excitement about it for several years, but mankind is nonetheless plunging headlong into the awesome, fearsome era of genetic engineering. Man, at last, is about to begin playing God.

What this means, in terms we can understand, is that human societies are now facing huge, unpredictable new challenges and, most likely, this world will never again be the same.

3

So, whether we like it or not, we are going to be dealing with the problems and the challenges of genetic engineering. Ready or not, genetic engineering is here; it's too late to even consider turning back. Progress is being made in laboratories all around the world, so no single nation has any way of shutting off or controlling or directing this new, continuing line of research. Obviously, there's no way to retrace all of those important little steps that have carried us—with the tools of modern biology—to the point of gaining some control over the stuff of life, the living genes. There's no way to retreat now to the warmth, comfort, and ignorance of the womb; no pretending this very brave new science —this new skill at tinkering with the blueprints of life—will suddenly just disappear. It's here, and it won't be forgotten or discarded. Genetic engineering won't go away, even though the people of this world remain totally, woefully unready to handle the consequences.

Basically, what we're facing is a true revolution, a revolution quietly spawned in biology laboratories, an insurrection that is gradually but certainly disenfranchising Mother Nature. It's a rebellion that's attempting to filch the sharpest, most dangerous weapons from creation's own armory. Scientists, indeed, are now picking the locks guarding some of life's most sacred inner secrets, and they're gambling that the information found may pay off someday in new products, processes, and life-styles we can't now even imagine.

Of course, there is very much yet to be done before biologists, biochemists, geneticists, and other artful specialists finally begin playing God, finally begin redesigning human beings. But that event—that strange possibility—may not really be as distant as some people believe—or would like to believe. Right now, scientists are already rearranging the genes of microscopic and submicroscopic creatures—the bacteria and viruses—to create new strains of tiny organisms that do strange new things. The remaining steps necessary before the same techniques can be applied to mammals—including man—are large and difficult, but not impossibly so. Look soon, indeed, for a biologist somewhere to announce he's finally built the first man-made mouse.

The work going on now with bacteria and viruses, too, may not seem terribly exciting, but some careful digging will show that even at this low level there is good reason for excitement. What it means, in one sense, is that scientists, and the rest of us, too, face the distinct danger that one laboratory worker somewhere, someday, will finally build the wrong bug; a disease organism that when released, even by accident, can't be stopped. Such a sorry possibility certainly supplies enough cause for concern, and many biologists and microbiologists close to this work are clearly worried.

Still another bit of perspective is important. Remember, first, that through all 4.6 billion years of the earth's history—and especially through the last 3 billion years of biological evolution—development of the various species has proceeded mostly through chance changes in genes, chemical changes called mutations. Most of these turned out to be duds, but slowly by trial and error and in response to environmental pressures, forms have evolved which take advantage of the small percentage of good mutations which occur.

Evolution, then, has brought us this far by repeatedly shuffling and reshuffling the genetic cards; scoring on good combinations, failing on others.

Now, however, with the tools of molecular biology and the purified sauces and syrups of biochemistry coming to hand, man proposes to take over Mother Nature's job of genetic card shuffling, with clear intentions of stacking the deck in his own favor. Whether he can do this job better, wasting less and avoiding serious heartbreak, remains to be seen.

One interesting way to look at the whole process of evolution is through the eyes of strict Darwinists, who have long contended that all the complex living organisms of this world are simply the storage sheds—the security vaults—for genes. In this view, genes, indeed, are everything; the reason for being, the reason for birth, for death, for life. Under this banner, evolution is concerned only with safeguarding, transmitting, and improving the sacred genetic material. After studying the workings of communal societies—such as those operated by ants, bees, termites, and man—sociobiologists can often point to specific situations where individuals are clearly, purposely sacrificed for the good of the whole society; probably sacrificed so that the gene pool—the genetic card-shuffling apparatus—can be perpetuated. Such individual sacrifice is what sociobiologists refer to as altruistic behavior.

Now, of course, we're seeing man's first feeble effort, through genetic engineering, to turn this whole system around, to make the genes serve individuals and their societies, rather than vice versa.

Scientists are starting out on this revolutionary path, though, in the face of the almost unbelievable fact that man—through those billions of years of near-random card shuffling—emerged and progressed to the point where he can think, wonder about, study, and even alter his universe, and then go on to ask that powerful question: "Where did I come from?" No other creature on earth is known to have that ability, nor to have need of it.

So here is man, a creature slowly built up from the collected, organized debris left over from burnt-out stars, striving now to understand, even dimly, a universe he sees ranging from the heart of the atom to the vast, dark expanses of space. Man, a complex, thinking product of the universe's continuing chemical evolution, is now proposing to gain control of those chemical and biological processes and finally start rebuilding the living parts of that universe to suit his own needs, whims, and fancies.

As we move into this challenging new era of genetic engineering, the main question that needs to be asked early in the game— but probably won't be asked until very late—is whether all the expected benefits from genetic engineering really outweigh the hazards. At this point nobody is able to assess completely whether the risks will finally balance out against what scientists hope will be achieved. Nonetheless, the specialists involved in research are going ahead, pointing out the following possible benefits:

• In medicine, scientists are hoping someday to be able to insert new genes—tailored bits of new genetic information—into defective cells in which the original genetic instructions are disastrously incomplete, scrambled, or missing altogether.

Such an achievement will give doctors a bold new approach to the treatment of more than a thousand strange diseases—the genetic diseases—which are for the most part now untreatable. Examples include hemophilia, known as the bleeders' disease; Tay-Sachs disease which produces mental retardation, blindness and early death among Ashkenazic Jewish males; and the more common disorder, diabetes, in which sugar metabolism malfunctions.

• In industry, genetic engineering is already riding a strong wave into the future, and this will grow. In some cases, companies are already designing custom-built bacteria to accomplish specific tasks. Such tailored microorganisms will someday be on the assembly line to produce important medicines, enzymes, and hormones, all of which are often difficult to produce by artificial means. As an example, scientists working for General Electric recently announced they've developed a new type of bacterium, an organism specifically designed to be efficient at gobbling up spilled oil.

The wine industry, too, will probably learn to use genetic engineering. Yeasts used in wine production could be manipulated to improve the speed of fermentation, the quality of the wine, and the cost of the process. Conceivably, alcohol-tolerant strains might be developed that would last longer in a vat of fermenting wine, allowing vintners to produce a whole new class of stronger wines.

• Farmers—and food consumers—would benefit enormously. Given the benefits of genetic manipulation, future crops will make even today's best bumper crops look puny and anemic compared to the yields, the superfoods that will be produced through cultivation of strange new types of vegetables, grains, and cereals.

Serious work is already under way, indeed, which researchers hope will soon lead to transfer of the nitrogen-fixing ability of the leguminous plants to other important food plants, thus passing on the ability of the legumes—the beans and alfalfas—to gather and make their own fertilizer directly from the air. Recipients of these useful genes—which enable the legumes to serve as hosts for nitrogen-fixing bacteria—would probably first be major food crops like corn, wheat, and rice, which now produce maximum yield only through massive use of expensive fertilizers. Such a development would surely benefit a world in which the majority of its citizens climbs into bed hungry every night.

Such lists of genetic engineering possibilities could run on and on, especially when it comes to discussing the ways one might want to alter the human body, the way we might choose to change the human approach to reproduction, or the way we'd like, someday, to be able to control rigidly the birth, development, and growth of individuals.

Such goals may sound repulsive, of course, when applied to the human animal, but in the field of animal husbandry we are already seeing widespread use of artificial insemination, sperm banks, and other techniques to increase production of high quality meat animals. Remember, too, that we've already heard, from England, of reported examples in which human eggs were fertilized in vitro —in the cold world of the test tube—and successfully implanted into the uteri of several women. These fertilized eggs reportedly grew and became healthy infants born at full term.

A similar approach, with cattle, is now being done almost routinely as a business.

In the case of cattle, several firms—one in Oklahoma, another in Colorado, and a third in Minnesota—have been carrying out commercial programs to increase the yield of good meat and dairy animals. The first step is to give a prize cow a hormone to make her superovulate: produce an abnormally large number of eggs. She is then artificially inseminated with carefully chosen sperm from a prize bull. Once fertilization has occurred, the cow's abdomen is slit open and the primitive embryos—called blastocysts—are removed. Then they are implanted into the uteri of waiting cows, which will carry these new prize calves on to term.

With such techniques in mind, one can ask how long it will be, really, before widespread use of such techniques will be seen in the human reproductive process? Indeed, how long will it be before we begin seeing newspaper advertisements about "wombs for rent," placed by young women who make a living by renting out their uteri as appropriate receptacles for healthy, fertilized ova, to carry the embryo to term and then deliver a pampered, healthy infant to its "real" mother. We'd have to ask, surely, about who is really the real mother; the woman who supplied the egg and paid the bill, or the woman whose warm, presumably healthy, and comfortable body provided all of the baby's nourishment and support?

Will we see, someday, a local registry of wombs for rent, an agency governing how these willing women are to treat their bodies, ruling how they will nourish themselves in preparation for commercial childbearing? Or will we bypass that altogether and go on to "growing up" our embryos in mechanical wombs as envisioned by Aldous Huxley in his *Brave New World?*

It should be mentioned, too, that such tricks of in vitro (testtube) fertilization and the transplanting of embryos are not what is strictly referred to as genetic engineering. They might best be called "cellular engineering," since they do not involve rearranging the genes themselves, but merely manipulating the developmental process. Nonetheless, they are often classed under the heading of genetic engineering because the phrase is so widely defined. In some instances, there are specialists who like to think of mere crossbreeding as one rather primitive approach to genetic engineering.

It shouldn't be long, however, before this type of cellular engineering evolves into real genetic engineering. It is one important step on the road toward the time when each fertilized egg will first have been inspected and even tampered with, improved in one way or another by insertion, deletion, or swapping of one or more genes. Certainly, with this in mind, the day may come when every egg cell and every sperm cell will undergo rigid inspection before being allowed to join and develop into a living human being. Beyond this, too, it seems probable that one day we will begin tailoring the genetic instructions: the blueprints, within these egg and sperm cells to meet certain standards.

Another possibility, unfortunately, is that we'll see foolish fads that bring us a gusher of freckle-faced red-headed boys, or a covey of svelte blonde girls who, genetically prepared, go through life never having to worry about being overweight, without ever being pimply or even awkward.

Perhaps, indeed, we'll end up making ourselves too perfect.

Added to these possibilities is the often-dreaded question of cloning. This has turned out to be the subject that catches the imagination, that titillates movie audiences and shows up so often in science fiction. The movie scenario generally involves some demented dictator—or mad scientist—who clones up a whole army of perfectly obedient, tenacious soldiers who mindlessly obey his every command. Or, in an alternative approach, the dictator who, to ensure his continued reign, clones up a few copies of himself to keep ready in case of the worst.

Cloning is just the use of single identical cells to produce a line of exactly duplicate mature individuals. In those cow-production procedures, for example, one could take that sixteen-cell blastocyst, separate those cells, and induce each to grow into a new individual. Each individual would have the exact genetic makeup of his fellow "womb-mates."

The key to cloning in animals is obtaining new, embryonic cells that haven't yet begun to differentiate into specialized cells. The goal is to find cells that haven't yet discovered they're to build eye cells, ear cells, toe cells, or nose cells. Once a line of cells has begun to differentiate, it becomes more difficult to induce them to become something else, or to tell them to start over again to build a whole new individual.

Cloning is going on now, and there's nothing even remotely sinister about it. Most of it is done to propagate special plant species. Orchid growers clone their most valuable varieties by taking a snip of tissue from the growing end—the tip—of a plant. This sample of tissue—known as meristem tissue—the tissue that hasn't yet differentiated into leaves, stems or flowers—supplies the individual cells that will be induced, through chemical signals, to grow and become whole adult plants almost identical to their single "parent" plant. In terms of genetics, the sets of genes that are hidden in the nucleus of each cell are identical, so these cells then produce identical plants. Thus orchid growers are finding this technique extremely valuable, since it lets them propagate endlessly their very best prize-winning plants. It also lets them escape the genetic lottery, giving them a way to reproduce plants without putting them through the hazards of breeding, flowering and seed-ing; without taking chances that a particular set of genes will be inconveniently reshuffled.

Another important advantage to cloning, biologists have noted, is that the meristem tissue is normally found to be free of infec-tious viruses. Geranium growers have been able to produce strik-ingly big and beautiful flowers simply by eliminating the burden of viruses found in most mature plant tissues.

There is, on the other hand, another side to this long, involved argument over genetic engineering and its value. There certainly are disadvantages—or at least potential disadvantages—involved, mainly because of the complex, almost unanswerable ethical prob-lems the science will open up.

Scientists involved in the work expect the most immediate areas of dispute to center on the current work with bacteria and viruses. There are certainly questions about the physical confinement of these new creatures that need to be answered, and questions about whether certain experiments ought to be done at all. But besides these are the deeper questions about whether the world's public should be placed in danger so that a few scientists can pursue their exciting work.

Although obviously important, and perhaps unanswerable, these questions were the subject of a large conference called to assess what the dangers really are and what might be done to respond to them. The conference was held at Asilomar, California, in February 1975. For the occasion, nearly 100 of the world's fore-most geneticists, biochemists, biologists, and microbiologists met with a few lawyers, government administrators, and a handful of science journalists to begin hashing out the issues. When it was all over, the scientists agreed, with some reservations, that there are some types of experiments that just should not be done, given present knowledge and conditions. Some other, slightly less hazardous work should be done only under conditions guarantee-ing maximum security to keep strange new organisms from escap-ing into the environment. The third category of experiments were those of far less hazard that can be done under present conditions.

They agreed first that some proposed experiments should not yet be done. These involved experiments like slipping the genes for cancer or smallpox or the botulism toxin into a widespread but normally harmless bacterium such as *E. coli*. Since *E. coli*—a normal, harmless, ubiquitous inhabitant of the human gut—is one of science's most important bacterial tools, researchers suggested it would be unwise to give bad genes such a convenient carrier for entry into the human body.

Another such experiment—while considered a little less hazardous—was also proscribed. This would involve the use of any human genetic material, taken from live human cells, being placed inside a virus or bacterium. These experiments were already being done in some laboratories, but conference members warned that the human genes might be carrying hidden cancer viruses. The danger in such experiments is that the virus, along with the human genes, could be transferred by mistake into an easily transmissible germ, and this might make the germ become a vector—or carrier—for cancer. This would act similar to the way a mosquito serves as the vector for transmission of malaria.

Another important danger not to be overlooked in this research is the possibility that the genes that provide bacteria with their resistance to antibiotic drugs might be even more widely, even more rapidly, spread around. Doctors have been discovering that even some of their newest, most potent antibiotics are becoming less useful because some bacteria seem, almost immediately, to become resistant. Worse, this resistance is spread to other bacteria —even to bacteria of different strains. Laboratory experiments have shown, indeed, that relatively friendly bacteria like *E. coli* can pick up these genetic resistance factors and pass them along by sexual mating to more dangerous bacteria.

The pressure to do this kind of research, however, is strong and constant. The biochemists, biologists, and others are working in a lively, challenging and competitive area of science, an area that seems sure to produce a number of future Nobel Prize winners. Right now, nobody wants to be left standing at the gate in this important race. The point is that so very much can be learned about life, about cancer and about how living organisms are controlled that it seems important to carry on with the work—but carefully.

Beyond this stage, this period of somewhat simple experiments with bacteria and viruses, will arise the deeper, more difficult questions, especially those involving ethics. It's not difficult to get general agreement about the value of producing more highly nutritious food from plants, or to agree about the benefits of developing stronger, better, healthier meat animals. But beyond this, things get considerably stickier.

The most difficult questions arise, predictably, when we start discussing how to tinker with the human set of genes, and especially when talk deals with making substantial changes. Nobody can say yet how substantial those changes may be, but the opportunities seem almost limitless. It has even been suggested that we may—in the distant future—want to build men and women who can live full-time underwater, like fish. This certainly would provide one powerful means of exploiting the sea—in itself a laudable goal—but we'd have to consider carefully how these people with gills might fit into existing societies. Chances are, of course, they might prefer *not* to fit into society.

Still, this whole notion that new types of human beings can be designed and built is the point where the most vivid imaginations take over. Suggestions have been made for redesigning the human backbone, to make it less vulnerable to slipped discs and other injuries, for making people much smaller—in order to make more room and cut down on food consumption—and also to extend the human lifespan considerably. Jokes here are rampant and sometimes misdirected. One Chicago biologist suggested in jest that an ideal goal would be to build little green men by adding chloroplasts—the cells plants use for photosynthesis—to the skin of humans. People, like plants, would draw their energy directly from the sun and thus do away with the bother of eating.

Unfortunately, this offhand suggestion was picked up in all seriousness by a writer who was doing an article for a national magazine. Now the poor professor is enshrined in history for making this awesome suggestion. All traces of the joke have thus disappeared, and little green men have become one of those probably mythical new things that people suspect will spring full-blown from genetic engineering.

But on a more basic level, a second look at the whole process of evolution can also help put the role—and the future—of genetic engineering into better perspective.

Evolution may be wasteful and slow, but one can't demean the success of this lotterylike system of gene shuffling that has brought man and his fellow creatures this far along. After all, evidence now in hand shows that the first tool-using hominid—the manlike creature who once lived in the wild territories of East Africa—probably began this slow evolutionary process toward modern man more than 3 million years ago.

This molten earth probably solidified and cooled, to begin forming an atmosphere, some 3.5 billion years ago, and life emerged in the form of bacterialike creatures, or something similar, probably about 3.2 billion years ago.

Even though life had finally emerged, however, scientists estimate it took another 2 billion years or more—years filled with repetitive episodes of random genetic experimentation, evolutionary maneuvering, and countless ecological dead ends—to produce a new, higher kind of cell, a cell that was fundamentally different from bacteria. It was a cell that contained numerous different components—including perhaps bacteria and other organisms—that had joined together to live in harmony inside that special enclosure. Indeed, some biologists think the energy-producing bodies seen today in living cells—the mitochondria—are the remains of early bacteria that found a comfortable place to live, in symbiosis, inside the cell.

The important point, however, is that it was from this single, more complex type of cell that all of the other higher organisms arose, and the first of these higher organisms finally appeared only 600 million to 1 billion years ago.

Once this process began, however, paleontologists believe it wasn't long before creatures made up of bones and muscle soon appeared. Evolution then brought on the reptiles, followed by the mammals, only about 100 million years before our own time.

As for man, about 2 million, 3 million or more years ago, a relatively small apelike mammal, named *Homo habilis* (tool-using man) by the late Dr. L. S. B. Leakey, gradually awoke to the revolutionary idea of making and using tools.

Of course, there's still very much yet to be argued about the times, rates, and routes by which all this occurred, but the feeling among paleontologists and their modern comrades, the anthropologists, is that the development of tools finally applied the pressure—evolutionary environmental pressure—that led to development of a better brain—a brain capable of making even better tools, including weapons. What it means, specialists suggest, is that such developments forced human evolution to move off in new directions which in the end resulted in development of our own species, *Homo sapiens*, only some 200,000 years ago.

It's not difficult to see, then, how closely development of the brain is tied to our species' increasingly rapid rate of evolution. Only 10,000 years ago, for example, man finally began to develop agriculture. About 4,000 years ago, he began keeping some records of his own history, and only 400 or 500 years ago did we first begin delving tentatively into the realm of science. Really serious efforts aimed at predicting the future began just 20 years ago, and now, with genetic engineering, we're proposing to control our own species' future, perhaps even avoiding what up until now has been the eventual, unavoidable fate of all species—extinction.

James Bonner, a respected member of the biology faculty at the California Institute of Technology, likes to point out: "As a biological aside, I would remind us all that the normal expectation of an animal species such as our own is to arise through mutation, evolution, and selection, and then to die out and be replaced by a species more fitted to the then current environment."

Bonner points out that mutation and evolution have already created—or invented—millions of species of very different organisms since those first living things arose in the seas 3.2 billion years ago. Almost all of these species, regardless of how tough and ferocious, have been discarded, have become extinct, have been replaced by hardier, more able, and versatile descendants.

What, then, makes man think he's any different? What makes us think we can opt for immortality—immortality as a species, anyway—and get away with it?

The brain, of course, is the key. Not only is man growing more and more conceited, he's also growing more and more versatile, and the brain lies at the root of this flexibility. As anthropologist Margaret Mead once pointed out, humans above all other creatures do have the enormous ability to learn rapidly; we are truly flexible.

Bonner, at Caltech, pointed out a striking example of such human flexibility when he recalled a man who went "from child of a headhunter cannibal to professor of mathematics in a major university in one generation. It's a true story."

Further, Bonner reasons that if individuals are so very flexible, there seems to be no strong reason why mankind can't move from being a species doomed to extinction to become a species facing a long, long tenure here on earth, and perhaps elsewhere. He contends, perhaps too optimistically, that we can change our genetic destiny within one or a few generations.

We shouldn't conclude, however, that just because man is now gaining control over his own genetic machinery that chance changes, mutations, and new combinations of genes aren't going to continue to be important. Genetic change is bound to continue.

Already, of course, some subtle and unannounced forms of genetic engineering—falling perhaps into the category of selective breeding—are being carried on as the result of pressures inherent in the societies we live in. In the primitive past, for example, individuals with poor eyesight had little chance of surviving all the way to reproductive age if they had to cope in the wild. Today we fit them with complex lenses that allow the handicapped to function as normal, procreating members of society. This means that the genes that give less than perfect eyesight are preserved, left alive in the human gene pool to be spread even more widely. So it seems that Mother Nature, once upon a time, had her own cruel but efficient way of eliminating poor genes. Now we've taken matters into our own hands, and the strength of the human gene pool can only be weakened. Thus we are slowly reengineering the

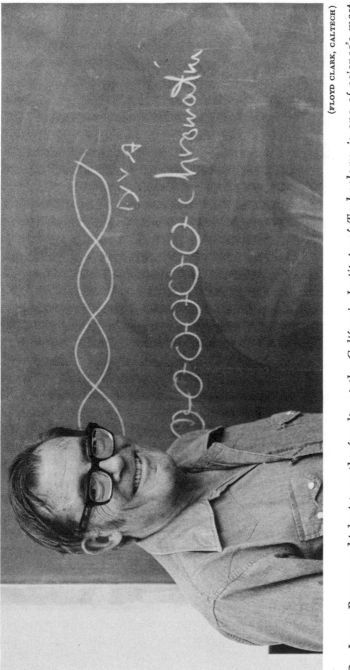

(FLOYD CLARK, CALTECH)

Dr. James Bonner, a biologist on the faculty at the California Institute of Technology, is one of science's most articulate enthusiasts for the rational development of genetic engineering. Bonner's forte, along with basic research on genetic materials, is the application of new scientific lessons to practical problems. He succeeded, for example, in greatly boosting Malaysian rubber production by treating the trunks of productive rubber trees with ethylene, a natural chemical.

human species, but not in a good way; actually undoing some of the painstaking work that evolution began deep in our past.

A recent, infamous and more overt approach toward the selective breeding of humans came at the hands of Germany's Adolf Hitler, who tried to force the issue in his favorite direction, hoping to build his superrace of fair-skinned, blond, tall, and lanky Aryans. His attempt to breed a pure race was, however, bound to fail, since there is no such thing as a wholly pure race. Breeding for beauty, for blondness and tall stature is a possibility in biological terms, but a difficult task now because of the taboos and preferences of the communities we inhabit.

Another experiment we are unconsciously doing with genetics is the result of economic, educational, and social pressures. More and more unusually bright children—some of them even geniuses —are being born as the brighter people of this world begin meeting, courting, and marrying on or near our university campuses. In brief, then, it becomes obvious that our value system—our emphasis on education and achievement through training—is beginning to carefully select the brightest people from within our population, is pooling their genes in a new subpopulation. This will possibly result someday in the gradual emergence of a new superintelligentsia, a new superelite.

With this in mind, one could ask, then, if this is already being done almost inadvertently through accelerated evolution, what need is there for genetic engineering? Can't we get new enough, different enough, talented enough people, plants, and animals through careful selective breeding without going to all the mess and bother of genetic engineering?

The answer is yes—up to a point. But evolution is slow, and our modern societies, if given the choice, usually opt for getting things done quickly, and hang the consequences. The motto, as that old-time labor leader Samuel Gompers put it, is, "More, now."

Still, many people will remain puzzled about why anyone would want to involve our own species, *Homo sapiens,* in genetic engineering experiments. They may easily agree that manipulation of the genes to create better, more productive, nutritious food crops is all right, and that new farm animals may be useful. But why fiddle with the human set of genes, the human genome?

The answers are numerous. First, as noted earlier, biologists, biochemists, and doctors would hope, through such fiddling, to someday eliminate the sad results they see coming regularly from birth defects, mongolism, and other rather common genetic afflictions.

Second, as this world becomes ever more crowded, and resources grow scarcer, the wisest societies—the societies with the will and the tools—are going to begin voluntarily limiting their own birth rates. One logical result of this, then, is that if each family is limited to two or fewer children, the thing to do is insure that the children they do have are the best possible children, born free of genetic defects. The best way to do this, they'll find, is probably by selecting or building up the best sets of genes possible, regardless of who the parents are.

This approach would certainly require some changes in attitude. That ancient, honorable idea of machismo, the notion that Real Men are obliged to spray their genes around as widely as possible, will have to be abandoned. Some measure of self-sacrifice —the sacrifice of one's own genetic heritage—will be necessary in some cases. Parents will still be able to have babies—if they choose to go through all of the pain and inconvenience—but they would first go down to some reproduction center to select the proper sperm and egg to get the characteristics they want in their children.

Another point, one of the things scientists, science fiction authors, and futurists like to discuss, is how miserably low the return is that we get from our investments in education. At present, for example, in many cases in the United States we spend some twenty or more years—and great bundles of money—educating persons who will each be productive for only forty or fifty years.

What this points to, of course, is that through the science of genetic engineering—through selection of genes that yield long, productive lives—it should be possible to decrease the length of time we spend being educated, and also greatly extend the amount of time we are productive. Some visionaries even foresee the day when the amount of time spent being productive and useful will be some 250 times longer than the amount of time spent being educated. Such a long working life may sound implausible, if not impossible, but if man is given enough control over both his genes and his environment, it still seems credible that our working years will be extended much beyond what they are now.

But conquering death, on the other hand, is quite another question. Death, many observers believe, will always be necessary if only to ensure that change can continue, that there will be room and resources left for improved, more versatile versions of the human animal. It's said that without death there would be no progress.

The goal, then, would not be to avoid aging and dying, but to make those final years more productive, alive, and healthy right up to the very end. A reasonable task would be to make those last years as strong and important—and as enjoyable—as possible by eliminating degenerative conditions such as rheumatism, arthritis, cancer, and heart disease.

Actually, it already seems logical that application of genetic engineering techniques to the infirmities of the aged might save as much time, heartbreak, and money as similar application of such techniques would to the unborn. It's well known, for instance, that a large percentage of available hospital beds hold elderly persons, many of whom have no prospects for ever getting well or living comfortably. Don't get the impression that this is a call for euthanasia for the elderly. It is a suggestion that the aged could somehow be better served, and that one of those ways of serving might eventually be through genetic engineering.

But genetic engineering for the sake of health and comfort is merely one thing. There is also the question of genetic engineering for the sake of harm, of using this new tool as a quiet weapon of war, a weapon of conquest. On this topic, certainly, there are whole gangs of pessimists and optimists, but if human history is any guide, genetic engineering will probably, somehow, be purposely misused sooner or later.

As one might expect, scientists watching this work believe that the most important, dangerous misuse of genetic engineering will probably come as a result of government efforts. It may not be wholly malicious, certainly, but the abuse will come dressed in the garb of "defense," or "preparedness," or just under that innocuous catch-all phrase, "research."

In a conversation at Northwestern University, Dr. Robert Gesteland, a member of the biology faculty, commented that government agencies "are always looking for ways to control people." This, he said, is reminiscent of early attempts to keep American Indians calm and contented—and very docile—by supplying them with alcohol. Now, Gesteland said, instead of trying to pacify Indians, the great numbers of people trapped in ghettos have become the target. They are even seen, perhaps unconsciously, as being somehow subhuman.

"The trouble," Gesteland said, "is that nobody knows how much human genetic engineering is already being done by the government. It would seem strange indeed if some efforts weren't being made in that direction.

"I think the prospects are very frightening, and I don't see any way to prevent it from happening. Nobody's going to ban restriction enzyme [one type of gene manipulation] experiments because it's the most important research method around now for understanding life. The only hope is that it won't work very well."

But some scientists suspect that it might work all too well, especially at a place like the United States' former biological warfare research center, Fort Detrick, Maryland. Bonner recalls that several years ago, before biological warfare research was publicly shut down, a rumor was circulating among biologists that their colleagues at Fort Detrick had succeeded in transplanting the genes for botulism into *E. coli*. Bonner said he wrote to the commander of Fort Detrick, telling him of the rumor and seeking an explanation. This letter went unanswered, so the Caltech scientist said he wrote another, stronger letter, warning of the harm that could come if such a rumor were made public.

"I got a nice letter back," Bonner recalls, "and he said it was not being done, nor was it even contemplated."

Suspicion, however, persists—on both sides of the Iron Curtain.

One point that should be made, too, is that creation of an unstoppable bacterial weapon can, under proper circumstances, be as dangerous for the developer as for the opponent. Unless extraordinary efforts are made in advance, or for defense, chances are the disease created may return to infect its creators.

Still, of more immediate concern to scientists in the field, is the growing fear, even among specialists, that a new lethal bug, designed solely for research use, will accidentally find its way out of the laboratory and get loose to begin infecting the human population.

While this does provide reason for concern, other scientists warn that too much concern, too much fear, is already leading to timid science, is causing a serious slowdown in vital areas of research. Stanford University's Dr. Joshua Lederberg, in his remarks published after that Asilomar Conference, put it this way:

> In our preoccupation with the risks of creating artificial diseases we may deny ourselves the tools to cope with the global, natural evolution of existing organisms. As the custodians of an ever more crowded planet, we must look to research on viruses as one of the keys to survival. Besides this, we should also be multiplying manyfold our often piteously small public health measures for global health. We must also keep in mind the paradox that a side effect of advanced hygiene and the prevention of disease is the emergence of whole populations of naïve hosts, protected since childhood from the experience of life-threatening infection, and in some cases—for that reason—even more vulnerable to new epidemics!

The message, of course, is to think, to think hard about the uses and abuses of genetic engineering; to begin erecting, in some way, the safeguards needed to meet this new challenge. Similarly, at the dawn of the Atomic Age, all we saw was the surface; the huge explosions and the quick winning of a terrible war. Yet today, after Hiroshima and Nagasaki have been rebuilt and the monuments erected, we're still fighting the battle over atomic power as we watch the nuclear disease spread around the world. Use of the atom turned out to be one way to end a war, but it may yet provide a way to enslave ourselves.

Will it be the same with genetic engineering? Are we setting out to lighten the burden of disease, of mental illness, age, cancer, and other infirmities in ways that will spawn strange new burdens we can't yet foresee?

Lederberg's colleague at Stanford, Dr. Paul Berg, said it best: "Now, and in the next few years, we shall be doing things that would have been thought improbable a few years ago. Genes from virtually any living organism can be put into another, completely unrelated, organism.

"And," he warned, "the time for reckoning may very well be today."

2

The Science

MENTION BIOLOGY and most people wrinkle noses in awful re-
membrance of things smelly, icky, gooey; of things wriggly,
creepy, and crawly turned loose in the pandemonium of a high
school laboratory. Up to a point, of course, they're right. Things
are still smelly, icky, wriggly and creepy in school laboratories.
Pandemonium still reigns because, after all, biology deals with
life in all its forms. That, of course, is what makes it so important.
That is what makes the topic of genetic engineering so awesome.

In some laboratories, especially those run by biochemists and
biophysicists, life often seems very remote and distant, far removed from the vials of chemicals, the strange machines, and the
impressive glassware. Nonetheless, biology as a whole is still
centered on life—on living, breathing creatures. Biologists are still
very interested in what goes into living things, in what comes out,
how life is organized, and how everything is controlled.

Most interest now, certainly, is beginning to focus on control
because control is central to everything. Without control there is
nothing; organization is impossible. Nothing alive could exist. So
in biology these days, control is where it all begins. Control is
crucial.

And control, of course, brings up the whole subject of genetic
engineering, which is essentially man's attempt to take command
of the biological controls, to take command of this fragile living
enterprise and, like a captured ship, steer it off in new directions.

At present, however, modern scientists have only now begun prying open the wheelhouse door aboard what might be called the biological ship of state. Acting rather like a band of imaginative pirates, they've now forced their way onto the bridge, but they haven't yet seized the wheel. But watch out—it's only a matter of time.

Indeed, the learning process in the biological sciences can be accurately likened to an old-time pirate crew's attempt to capture and control a modern ocean liner. As pirates, scientists have successfully stopped the vessel, have gone aboard, and have taken command of her decks. Now, they're trying to learn how she runs.

So far, they've pretty well determined her general shape, what fuel she runs on, how she's handled and, alas, how to sink her. What they still need to learn is the hard stuff: how does she navigate, how was she built, how does she repair herself, and where do you get spare parts? They're also largely wandering in the dark when it comes to actually building such a vessel; yet they're getting wiser all the time, and they've got the capturing part of the business down pat.

So, too, in actual biology. It has taken many years, but scientists have been able to capture and closely study living creatures and they have built up a collection of living things ranging from the lowliest virus all the way up to that most complex example of the mammals, the thinking animal, man. Finally scientists can ably and accurately describe their specimens. They can measure them, run them through tests, dissect them, weigh them, even kill them. They can even begin building living things such as bacteria and viruses, but they're still unable to repair higher organisms or supply important new parts that work.

This picture is changing fast. Through years of careful work, countless dead ends and frequent restarts, the world's biologists have worked themselves ever closer to finding out, really, what it is that produces the unexplained commodity called life. It has long been known that the living cell is, for most creatures, the smallest viable unit, the last, the smallest indivisible piece. If they try to break life down further, life ends. Thus you can break up the body, disconnect some cells from each other and, under proper circumstances, these cells can be induced to live on. But break up the cells, and that's it.

Biologists, of course, haven't let that little barrier stop them. The learning process has taken experimenters much beyond this, and scientists have learned a great deal about what's hidden inside the living cell, how important some of the things found inside are, and which are the most—or least—expendable. Actually, it has been well determined that the living cell is built, and operates, like a finely tuned machine. Given the right conditions and the right support, it can run without much danger of failure and, better yet, it can actually repair itself. Mistreat it, however, and the end comes quickly.

These cells, these machines, come in all sizes, shapes, and colors. But more important, cells come in various types which are carefully programmed to do different things. Cells in higher animals are specialized; special cells with special talents are carefully grown to do specific jobs, to perform different tasks according to where they find themselves in the body. And it's such specialization that's important for progress in life, and in the study of life.

Obviously, nose cells build noses, not livers. Skin cells make and maintain the body's protective covering, while refraining from making things like new bones, tongues, and teeth.

This talent, this ability to make specific cells for specific jobs, has become one of the important and challenging problems of biology, and scientists have been struggling to learn, in detail, how this is accomplished, how it is controlled. How, they'd like to know, does a single cell—starting from a very generalized, undifferentiated state—build up specialized new cell lines while dividing and redividing, over and over? How does one and only one line of these new daughter cells know enough to become parts of the visual system, some making eyeballs, the others making eyelids, others the optic nerves, and still others the brain's optical cortex? Where is all this information stored? How is it processed, how is it expressed, how is it transmitted?

Such questions, certainly, revolve around the problem of control. What is the control mechanism that builds, organizes, and runs the living body? What is the source of biological information

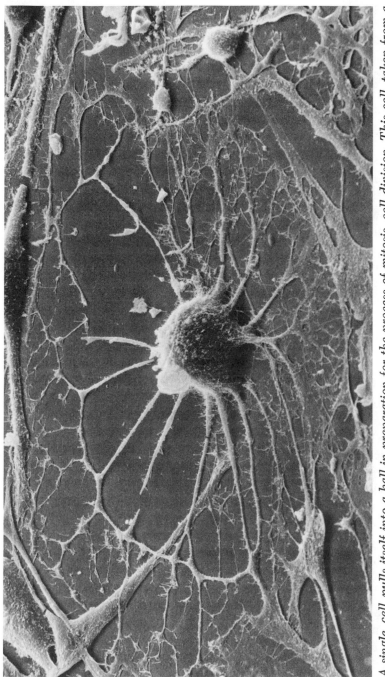

A single cell pulls itself into a ball in preparation for the process of mitosis, cell division. This cell, taken from a mouse, was prepared for viewing in the scanning electron microscope by Dr. Jean-Paul Revel at the California Institute of Technology. Magnified 400 times.

that tells a growing baby—first while inside and later outside the womb—to grow blue eyes, brown hair, a pug nose, and how to mature on schedule?

For the sake of brevity, just blame the genes, even though this isn't nearly complete or complex enough an answer. The whole process of development is so very complex that it requires several disciplines within biology—genetics and embryology, for example —to try to understand it. Perhaps the simplest approach for the nonscientist, though, is to view the genes as the body's chemical blueprints—or even as the central planning agency—directing how the living creature is built. Genes supply the details, the scheduling, and the commands for building and running the whole organism.

Added to this, of course, is the whole question of heredity. The genes also carry all of the odd and interesting characteristics that make a family unique. The genes act as the repository of the past, the chemical storehouses where all that is good—or bad—in a family's biological background is kept. The genes carry the information that is passed on from the parents and joined in the building of a new child.

What, then, are genes?

That's not simple, either, but the clearest approach is to say they're tiny—submicroscopic—bits of information that are stored in the form of precisely arranged and rigidly controlled chemical signals, or codes. It is from this store of signals or blueprints—read and acted upon inside each living cell—that the cells learn how and when to do their tasks for the sake of the whole body.

A good analogy would be to view the genes as the music score for a whole symphony orchestra—except that the genes do even more. Not only would the genes, like the music, specify when, how long, and how loud each instrument should be played, they would also specify how each instrument was to be built, where each would be placed in the orchestra, how each was to be tuned, and how long each would last before wearing out.

So, again, this raises questions of control. Biologists have to ask how all of these specialized jobs get taken care of at just the right times, in the right places, and in the proper amounts.

A partial answer comes from one important fact that biologists learned some years ago. A crucial fact in terms of what scientists hope to do with genetic engineering in the future: each cell—like each member of that orchestra—has a full set of genes, a complete copy of the score.

Too, each cell, like each instrument, plays—by itself—only a small portion of the whole program. All together, however, the result is a magnificent, triumphant, living, breathing, lively creature.

When compared to any one instrument, however, the cell's potential versatility far outweighs whatever an oboe or violin might be capable of. A given instrument certainly isn't able to produce every sound of the whole orchestra. Each cell, however, can pull from its own store of information—under the right conditions—all the instructions needed for reproducing the entire body, or any part of it. We see this ability perhaps most readily in plants, where a single cell can be induced to restart, can be stimulated to build a whole new plant.

One of the important factors behind this talent, scientists have found, is the cell's ability to sense and interpret the environment it finds itself in. Plant cuttings, for instance, sense, when stuck into moist ground, that some of those undifferentiated cells hidden in the stem must change quickly and become root cells. Similarly, a potato, when sliced in half, immediately begins responding. The cells on the exposed surface start up to build a new protective covering of skin. Likewise, if a lizard loses his tail to a predator, then the cells adjacent to the open wound soon begin building a new tail. This illustrates, of course, that all of the genetic information is readily available, and that it can be called up and used quickly when the need arises. Through genetic studies, then, scientists hope to find out what the environmental signals are, how to duplicate them, and how to turn cells on and off by using artificial signals. Man and the other higher animals have lost this "restarting" ability through evolution, but we may yet regain it through genetic engineering.

Scientists are beginning to strongly suspect that this spectacular ability for cells to start over may be what occurs by accident in cancer. It seems probable that sometimes the control systems of one or more cells fail, and thus a specialized cell somewhere, say in the skin, is suddenly rudely awakened from its humdrum, specialized role. With its control system suddenly excited, or overstimulated, the cell misreads the signals it gets from its neighbors and starts playing the whole score, saying to itself, in essence: "Hey, I'm the first cell of a brand-new baby! I've got to get busy now and build a whole new person, fast."

And thus the cell begins dividing rapidly, producing lots of new daughter cells, trying to build a whole new person. These new daughter cells, too, appear to inherit their parent's upset control system and so exclaim, in unison, that they, too, must each build a brand-new individual. They all keep dividing rapidly, wildly out of control. The trouble arises when they begin colliding with the formerly healthy organs already occupying that spot, and the tumor begins shoving other tissues aside as well as gobbling up most of the nutrients. The new, vigorously growing tumor, containing thousands—even millions—of cells that are each trying to build a whole new body, becomes a grotesque, deadly cancer that eventually destroys the living body.

The sad part, besides the death of the individual, is that for both the original body and the growing cancer, this frantic attempt to build more bodies is strictly a "no win" situation, a biological dead end. The effort kills the host; the growing tumor then loses the source of its support, and so also dies.

It should be noted, however, that biological accidents like this don't really happen very often, despite our fears of cancer and the disturbing statistics. Through the information already stored in the genes, the living body actually expects this to happen occasionally, even in a healthy individual, and has already set up control systems designed to handle such mavericks. Defense mechanisms—and, most importantly, the cellular immune response —normally mobilize quickly to destroy aberrant cells almost as soon as they begin running out of control, early enough to avoid the danger. The existence of such surveillance systems implies that persons unlucky enough to get cancer may be suffering first from a weakened or defective defense system. This cellular immune system, too, is the strong system which blocks implantation of new organs, since it can recognize foreign tissues as "nonself."

Aside from the problems of cancer, each cell's having a full set of the body's master blueprints makes the contemplation of genetic engineering possible. It leads, indeed, to ideas like cloning, or asexual reproduction, and the possibility of building healthy new spare parts for the body. One day, when biologists and other scientists learn to take the genetic controls fully into their own hands, it will probably become possible to snip a sample of tissue —a sample of healthy cells—out of a patient, place them in a petri dish and add the right chemical instructions that order these cells to grow up a new heart, lung, or kidney that can be retransplanted into the patient. This new organ, of course, would be made up of tissue that originally came from the patient, so the powerful cellular immune system would still recognize the new organ as "self" and be less likely to reject it as a foreign invader.

Before tailor-made new organs are available, however, most specialists expect medical scientists to make an end-run around the tissue-rejection problem. Instead of growing up a whole new organ as a means of avoiding rejection, scientists are working now to find ways for selectively shutting off the immune system so the body can accept a "nonself" organ from another donor.

No one can say when either process will see use, but a good illustration of how close we've already come to the time when cells can be ordered to do new things is shown by the work of a team of biologists at the California Institute of Technology, a team headed by Dr. James Bonner. This work is still in its infancy, but Bonner reports that sets of fully differentiated cells—cells genetically programmed to live and work in the liver—have been induced, by new outside chemical signals, to stop producing liver proteins and begin, instead, producing a protein found only in the brain. The results suggest that scientists may finally be learning what some of the chemical signals are that turn genes on and off. At the same time they're also gathering hints about what the signals are that produce other specific responses from the genetic blueprints. They know all the information is there, well hidden inside the cell's tiny nucleus. Perhaps all that's needed now is to work out the correct signals that will bring different genes into use at will.

Interesting proof that each cell contains a full set of genetic instructions, and that these instructions can be reactivated even when the cell is "dormant," came from the work of Dr. John Gurdon, in England. In an elegant experiment years ago, Gurdon took an unfertilized egg from a toad, known in the trade as *Xenopus laevis,* and tricked that egg into building a complete new toad!

Gurdon scooped a complete nucleus—containing a complete set of the toad's genes—out of a differentiated cell taken from the lining of the intestine; a real dead-end cell. At that time, these genes were programmed to make nothing more than the products of the intestinal lining. Then he inserted this nucleus into that unfertilized egg after the egg's own original supply of genes had been carefully eliminated.

The result, in time, was production of a healthy young toad that grew into a normal adult, a carbon copy of the toad that had donated the set of genes from its intestinal cell. Gurdon thus proved that the genes in any cell are capable of building the entire organism, and that under the right circumstances they can be turned on again and put to work.

Such proper circumstances, then, include the necessary signals from the cell itself; and the signals that the cell reads from the environment around it. Thus the cell says to its own nucleus—the little inside domain that contains the chromosomes, the genetic materials—something like: "I see we're supposed to build a new toad. Get busy."

At a later stage, when much more tissue has been grown and the cells begin specializing to handle specific jobs, different cells begin turning into eyes, tongue, and other organs. Other cells tell their nuclei, through more chemical messages: "We see we're at the end of what will become a leg. Let's start building toes."

The same procedure, of course, applies when the organism is complete. The cells, sensing what has happened, tell their nuclei: "We see there are enough toes, big enough toes, now. Stop."

This, again, brings up the problem of cancer, which seems so intimately related to the body's processes of growth and control. We now strongly suspect that malignant cancer cells have lost the ability to signal that important word, "Stop." In more technical terms, biologists speak of a cell or group of cells as having lost the property of "contact inhibition." This means that cancer cells, unlike normal cells, don't stop dividing and building more tissue when they run out of growing room. Cancer cells keep on dividing madly, growing out of control until cells pile up on top of each other in an ugly, chaotic mass. This mass, of course, is the tumor.

And again, the problem is control. Biologists are moving closer to gaining control, and the prospects are awesome. Let's look at cloning.

Gurdon's work with the toad, *Xenopus laevis,* represents a most important step toward that goal of cloning animals. While the toad, in many respects, isn't as complicated as the human animal or other mammals, it still shares many of the same basic characteristics—and most of the biochemistry—with higher animals. Gurdon's work shows, startlingly, that it should be possible to take tens, hundreds or even thousands of identical cells, all containing the same genetic blueprints, and induce them to grow up into tens, hundreds, or thousands of identical copies of one parent. Extending this to the logical extreme, it should eventually become possible, if the world can stand it, to grow up thousands of copies of one individual: an Albert Einstein, a Henry Kissinger, or a Martin Luther King, Jr. This, of course, is a long leap from the asexual reproduction of one small toad, but it does point to what the possibilities are.

As for practical applications, cloning has long been a valuable tool in agriculture, and it will also become important in animal husbandry. Dairy farmers, cattlemen, poultrymen, and veterinarians all recognize that some method for perpetuating a particular set of desirable genes would revolutionize the meat and dairy industries.

As things are now, hefty stud fees are still being paid for the services of a prize bull, although frozen bull semen is now routinely being flown all around the world in a continuing effort to improve breeding stocks through artificial insemination. Unfortunately, this still represents crossbreeding, so the genes are still being reshuffled.

The question that needs asking, though, is what happens when, one of these days, the perfect cow, the perfect bull, is born? How do you go about preserving that ideal set of genes? If you mate this perfect animal with another, you risk rearranging that set of genes in their offspring, redistributing that valuable set of characteristics. Worse, if you fail to breed that perfect animal at all, you'll not only lose the whole set of genes when the animal dies, but you also won't have preserved any of them by passing even a portion on to another animal. By mating you can at least preserve portions of that set of genes.

Of course this brings up one of the most interesting facets in the study of genetics. Each cell normally carries two copies of each chromosome, one of each contributed by each parent. This gives every individual what is known as a diploid, or double set of chromosomes.

Therefore each cell also gets two copies of every gene, since the chromosomes are the structures—as seen in the microscope— on which individual genes are carried. In many instances, however, only one gene out of such a pair is expressed, meaning that one gives all the orders and the other just comes along for the ride. It's important, nonetheless, to have both sets, even though in some cases characteristics are controlled by a dominant gene while the recessive gene seems to just sit idle. In other instances both genes are seen to be at work.

Eye color, for example, is such a case of dominance. If a person receives two genes—one from each of his parents—coding for brown eyes, he can have only brown eyes. Furthermore, he can pass on the gene for only brown eyes to his own children. In a different case, if a person gets one gene for brown eyes from one parent and a gene for blue eyes from the other, he will still have brown eyes, but some of his children have a chance—provided he marries someone who also carries a gene for blue eyes—of being born with blue eyes. What this says too, is that a person who inherits blue eyes must—through the genetic lottery—get only blue-eye genes from his parents. Slip one brown-eye gene in, and he will have brown eyes. This means that the gene for brown eyes is dominant, that its presence commands the building of brown eyes, whether there's a blue-eye gene present or not. This, then, would mean that the genes for blue eyes are what is called recessive.

In such a system, then, if a person receives a dominant gene for any trait, then that is the gene which is expressed. Only when a person receives two recessive genes for a single trait can that trait be expressed in its recessive form.

In most cases—especially in instances where innocuous qualities like eye color, hair color, and similar unimportant characteristics are involved—expression of a recessive gene isn't very remarkable. In some rarer cases, however, a recessive gene may carry a serious defect which is normally masked by the presence of a dominant gene. Only when two of these recessive genes get together can the defect appear, sometimes fatally. Thus one might say that the system of dominance often provides a useful way to avoid the expression of some dangerous recessive genes.

This, indeed, is the main reason why matings between close relatives, such as brother and sister, are usually discouraged in most societies. Long experience has taught even primitive cultures that the chances of unfortunate birth defects increase markedly in close intrafamily pairings. Chances are much greater that both brother and sister will carry a family's bad genes for one trait, masked by a dominant gene. But match these genes up, and the odds increase dramatically that some of their children will express the recessive gene.

One should be warned, however, that the system of dominance and recessiveness is much more complex than the description here. In some instances, both the dominant and recessive genes are expressed to some degree, with the result being a trait that is controlled by neither a dominant nor a recessive gene alone.

In some instances, gene expression in one area may be linked, or be dependent upon, more than one gene. A good example is an albino person, someone who may have the genes specifying eye color, hair color, or even colored skin, but who has no color at all. Despite the coding for color, a defective gene for production of melanin, the pigment that makes eyes brown and skin black, may mean that no color is made in the body. Thus the individual emerges white, even though he may have the genes specifying black skin.

The same, of course, is true for eye and hair color, since the same biochemical system that makes skin color also builds the pigment for hair and eyes. Thus, if there is no color made, the genes can order up whatever they want, but they still can't build color in if the chemicals for color don't exist.

Similar, but with important differences, is the phenomenon known as sex linkage. Biologists have long known that a female gets two X chromosomes from her parents, while a male gets an X chromosome from his mother and a Y chromosome from his father. Indeed, receiving an XX set of chromosomes mandates that the embryo will be female. Given an XY set, the embryo becomes a boy.

This is an important system, since it provides a strong, orderly way for determining what sex the new baby will be, but it also has the potential for problems. Some genetic diseases such as hemophilia—the bleeders' disease—are what are called sex linked, meaning that they tend to show up in one sex, but not the other. Hemophilia, as an example, occurs only in males.

This occurs because the gene controlling chemical factors important in blood clotting is carried on the X chromosome. In normal people, this gene functions as it should, both in males and females. In hemophilia cases, however, the gene on an X chromosome is defective—a mutant—and the body fails to make clotting factors that help seal up wounds. Thus the men and boys with this bad gene on the X chromosome tend to bleed excessively even from what would normally be the smallest, most insignificant cuts. In women, the bad gene on an X chromosome is usually masked or hidden by a good copy of the gene on the other X chromosome, so there is no problem with the production of blood-clotting factors. But in men, the single X chromosome is masked by nothing, so the gene is expressed. In the rare cases where a female is dealt these recessive genes on both X chromosomes, she doesn't survive much beyond birth, if that long.

It's interesting to note, too, that this system allows a woman to be an unwitting carrier of this bad gene, and the danger is she will pass it along to her own sons and daughters. Some of the sons will get that defective X chromosome, and some of the daughters will themselves become carriers of the disease. As for the boys with hemophilia, all of their daughters will inherit that defective X chromosome, but none of their sons will—unless the father has the bad luck to marry a woman who is a carrier.

The history of this relatively rare disease, too, seems interesting. It is known, widely, as the "disease of kings" because of the frequency with which it occurred in males of the royal families of Europe. One version of the story has the gene originating in one of the royal houses of England, being passed all around the mighty families of Europe as the political marriages sent sons and daughters off to marry in other kingdoms. Actually, however, the gene for hemophilia has been known since biblical times, since it is mentioned as a disease in the Talmud.

But let's move on. Another important area of exploration is the whole process of cell division, since it is through this process that bodies are built and many genes are expressed. Cell division involves the splitting of individual cells into two new daughter cells, which themselves often go on to split into additional daughter cells. When the body's most common type of cells—somatic cells—divide, important, precise mechanisms come into play to insure that each of these new daughter cells gets a complete set of genes, an entire set of the body's blueprints. Since half of a creature's genes were contributed by its mother, the other half by its father, it means again that there are two copies of every chromosome and of every gene in every cell.

During this delicate process of division, all the structures—including the chromosomes—in the cell's interior go through a complex set of changes. Carried inside the cell's nucleus, the chromosomes move into specific positions where they split and new copies are made. This occurs because the split chromosomes provide a specific chemical template on which to build the new copies. Once the genes have been copied, the cell itself begins splitting and half of the now-doubled set of chromosomes goes to one daughter cell, half to the other. Thus each new daughter cell gets a full set of chromosomes, a complete set of genes. Too, each new daughter cell is usually an exact copy of its parent, unless some sort of biological accident has occurred that scrambles the results.

Biologists refer to this entire process—this exact dancing and splitting of the chromosomes that occurs during division—as mitosis. This splitting of individual cells into two more cells represents the way most organisms grow, and among the single-cell creatures like bacteria it is also the way they reproduce. Starting from one single lonely cell, the higher creatures' cells keep dividing until the number reaches billions.

There is, however, one important exception to the rule that each cell gets a full and complete diploid set of chromosomes. This exception is encountered during the process which prepares some special cells for the adventure of sexual reproduction. Specifically, this breach of the rules is found in the building of sperm cells in males and egg cells in females. After mating, these cells will get together to form a whole new individual. Naturally enough, since sperm and egg each contribute half of the new individual's genes, each of these cells must be produced in a way that insures they will only carry a half-set of genes. Thus a very special subsystem has been set up by nature to take care of this problem.

This halving of the gene set occurs in a process related to—but distinct from—mitosis, the normal process of cell division. Production of the haploid set—the half-set of chromosomes—in the egg and sperm occurs in a delicate maneuver called meiosis. The offspring of a union of two haploid sets of genes, then, ends up with a mixed bag: a dose of half its genes from its mother, the other half from its father. What it also ends up with is a set of genes that have been carefully reshuffled in what biologists refer to as the genetic lottery. The outcome of this gamble is based partly on luck, since the offspring has the chance of being made up of one of several different combinations of its parents' genes.

It should be remembered, too, that such reshuffling of the genes is very important in nature. This genetic lottery, indeed, has pro-

vided the powerful method—through evolution—by which creatures have constantly progressed, generation after generation,
toward more harmony with their environment. It has also provided the crucial means for coping with a changeable environment.

The result, gradually, has been populations of stronger, more
versatile—and in the case of *Homo sapiens,* more intelligent—
organisms. That's why, in every mating, the mother and father
each contribute half the genes, allowing Mother Nature to constantly search out fresh, improved new combinations.

Improved combinations, of course, are what plant and animal
breeders are searching for in their own crossbreeding experiments.
Already, in the few short years since the basic laws of heredity
were discovered, geneticists and breeders have improved food
crops enormously. Corn production has boomed—certainly because of better fertilizers, more efficient farming methods and
more knowledge about basic agriculture—but also, clearly, the
biggest boost in production has come through crossbreeding. Recently introduced, too, are some new corn varieties that contain
vastly increased amounts of one nutrient—an amino acid called
lysine. When in widespread use, especially in Latin American
countries where corn is an important staple, high-lysine corn
promises to help overcome some of the world's most serious
nutritional problems.

Similar gains are being made with other crops. Greatly increased yields of wheat can be expected as American grain companies begin introducing new strains of wheat that produce as
much as 15 or 20 percent more grain per acre than present
varieties. Also coming into use is a grain called triticale, a cross
between rye and wheat which is supposed to be slightly more
nutritious, and which is also expected to grow well in some mar-
ginal agricultural areas, like eastern Africa.

It is becoming increasingly evident, however, that much of the
easiest crossbreeding work has already been done, that it is now
becoming more difficult and expensive to find new combinations
of food-crop genes that are useful. Thus the next big step, prob-

ably, is going to involve genetic engineering, the actual building of new genetic combinations through manipulation of the genes themselves.

Scientists can now contemplate this possibility—for both plants and animals—because in recent years they've been learning more and more about how to tinker with and control the tiny organic molecules that make up the genes. It was learned rather early, for example, that the genetic material itself, the huge, complex molecule that carries the chemical codes, is a substance called DNA, an abbreviation for deoxyribonucleic acid. This important molecule—and a related molecule called RNA (for ribonucleic acid)—are made up of small chemical bases, or building blocks, called nucleotides, which fit together like the links in a chain. Only five of these nucleotides are normally found in genetic materials. Two of them are called purines, and three are known as pyrimidines. The two purines are adenine and guanine, while the pyrimidines are cytosine, thymine, and uracil. Thymine is found in DNA, while uracil is found in RNA.

In the cells of higher animals, DNA, the molecule of heredity, is commonly found inside the exclusive little enclosure known as the nucleus. In lower creatures like bacteria, there is no nucleus, so the DNA floats around freely inside the cell. DNA is also usually found in combination with specific proteins, and together they make up those rodlike objects in the nucleus, the chromosomes.

The big DNA molecule itself is made up of these nucleotides hooked together, strung together rather like beads on a string or links in a chain. They are hooked together by connectors called phosphodiester bonds. In a DNA molecule, two such strands, or strings of nucleotide beads, are found lying side by side, and at each bead, or each nucleotide, the two chains are connected by what are known as hydrogen bonds. In principle, then, the DNA molecule might be said to resemble a ladder. The sides would be represented by the two parallel strings of nucleotides, while the hydrogen bonds would be the rungs holding them together. It would also be somewhat analogous to a railroad track, with ties holding the two long rails together, parallel.

But there's a hitch. Each of those chemical bases—each bead on the string—will connect with only one other type of base on the other string; an adenine base on one side of the ladder will share a rung, or pair only, with a thymine base on the other side. A guanine base on one side, too, will pair only with a cytosine base on the other strand.

For purposes of illustration, then, the whole system might better be seen as resembling a zipper in which any given tooth on one side will pair with only one kind of tooth on the other, and none other. Thus, if you're going to build one side of a new zipper to match with another half you already have, in this case you'd have to know exactly what kinds of teeth were where, and then build your new half-zipper to match that order exactly.

This, of course, would represent an awkward, expensive way to build zippers, but for Mother Nature it represents an amazingly reliable and stable way to build and preserve the blueprints needed for making a living individual. In operation, she can unzip the molecule, then make exact copies of one side by using its mate as a chemical template. By this method, during cell division, she builds complete sets of the original genes, and each cell normally ends up with an accurate, complete set of genetic instructions.

In operation, too, the cell—when called on to make proteins like insulin—uses this same mechanism. Biologists believe that a portion of the chromosome bearing the needed information unzips so that the cell's copying machinery can make a single-strand RNA copy of the DNA's chemical code. Then the RNA is carried away from the DNA to be copied again. In this second copying step, the cell brings up the right amino acids—the building blocks of proteins—and hooks them together in the right order as instructed by the RNA. Thus the RNA acts rather like a tape recording of the DNA's instructions, telling the cell's protein-building apparatus how to make a specific product. Then, when enough of that product has been manufactured, somehow the DNA is informed and it stops making RNA copies of itself, probably by zipping itself up again to hide the instructions. Production stops abruptly.

As for the genes themselves, they are merely specific lengths of the DNA molecule, divided into various sizes along the strand by what scientists are convinced are biological start and stop signs. The difference between what two genes do, what they code for, is found both in their length and in the specific arrangement of their basic nucleotides. One gene, for example, may have an AAAGGC sequence on one strand, while another could line up like AGGGCA for a different function.

This mechanism provides thousands of different combinations, without duplication, giving nature the very wide variety of products she needs to build all the different types of living things. It should be noted too that evidence has been found suggesting that many copies of some genes exist in the chromosomes, making it possible for cells to begin production of specific proteins quickly, pouring out large amounts, once the chemical command is given. Too, the system can obviously be shut down quickly on demand.

The genetic code itself is what is known as a triplet code, since it requires three nucleotides on a DNA molecule to command which amino acid will be used at a particular spot in a protein molecule. For each amino acid the cell wants added to a certain protein, three bases—in exactly the right spots—are needed on the chromosome. Thus if a protein is made up of hundreds of amino acid bases, the gene needed to specify its construction must carry three times as many bases—the chemical codes—as the protein has amino acid links in its chain. In addition, each gene also needs those important start and stop signals to tell it where codes begin and end. What these amount to, perhaps, is a sophisticated form of genetic punctuation.

If we could easily see these long, ladderlike molecules known as DNA, they would appear most of the time as tightly coiled chains forming the long chromosome. In the nucleus they are matched, or tied up, with certain important proteins which apparently help keep them in the proper shape. In the normal configuration, the ladderlike DNA molecule is first twisted, rather like a length of rope is twisted, then wrapped in tight coils, or supercoiled, to resemble the shape of a corkscrew. Indeed, the chromosome, if stretched out completely, would be very, very long, and it might tend to tangle easily inside the tiny nucleus.

By this coiling and supercoiling the chromosomes are brought down to manageable size that can be used by the cell's chemical machinery.

It is also thought by some scientists that the proteins wrapped up with the genes may be part of the basic control mechanism that tells each cell what it is to be, what product it is to produce. These proteins could be the devices used for turning on some specific genes while keeping the rest turned off. Indeed, without this ability to keep most of the genes turned off, a person might end up with nose cells producing taste buds in the wrong places, or toenails, or hair growing on the bottom of his feet. Thus control, again, is terribly important, and suggestions are that control of the genes may be exerted through the action of these special proteins found in the chromosomes. It has been suggested, too, that this control mechanism may involve a simple act like having one of these big protein molecules actually sit on a portion of the DNA, keeping the DNA from being read by the cell's protein-building machinery. So, when a new gene product is needed—a product made by the gene that the protein is sitting on—something, some message from the cell or even from outside the cell, comes along, dislodges that big piece of protein, and allows reading of the gene's chemical code to begin. When the job is done— when enough of that gene's product has been made—then the signals from outside might cease and the big protein would then hop back aboard, blocking further reading of the code.

Another approach suggested for this same result would have the proper section of DNA, the part to be read, uncoil slightly, loop outside the protective protein coat and be read by the cell's production system. Still, in either approach, it remains important for the DNA to become available for reading while the rest is carefully kept out of action.

Perhaps the remarkable thing about this whole gene-reading system, once again, is its stability and reliability. Even though it seems somewhat complex, it does provide a surefire method— most of the time—for preserving the integrity of the genes, for insuring they remain almost the same down through generation after generation.

Still, when it comes down to the problems of genetic engineering, what scientists are most interested in is change rather than stability. In the process of evolution, Mother Nature has systematically taken advantage of small changes in the genes, using such changes to improve the viability of species and help them adapt to a changing environment. Nonetheless, even though nature is quick to take advantage of good genetic changes—all changes in genes are called mutations—the great majority of changes that do occur in genes turn out to be bad. Most bad mutations, making individuals less able to compete in their environments, are soon washed out of the population. For modern humans, such innocuous features like a family's hair color or eye color may not present problems if changes occur; but for creatures in the wild, even the smallest change can make a significant difference. A slight change in hair color for a creature relying on color and pattern for camouflage, for example, could mean he might be more easily seen, more easily selected out of the population by predators. Being killed early, too, would probably mean he was unable to produce as many offspring as normal. Thus, fewer members of the population would end up carrying that mutant gene, and it might soon disappear.

In rare instances, however, a mutation comes along—again, such as a change in color—that enhances an individual's chances of surviving, such as making him slightly better camouflaged than his brethren, permitting him to live long, breed more, and pass along this useful new trait to his offspring. If the trait is very beneficial, it soon enters the population widely and is thoroughly spread within several generations. The individuals which are products of mutations and show such changes in their genetic heritage are called mutants. They are—in either obvious or hidden ways—different from the other members of the species from which they spring.

This fact comes in handy in plant nurseries, for instance, where an alert grower is always looking out for any new, different-appearing plant that shows up in a field or orchard. If one plant stands out because of its different color, shape, size, or quality of fruit, chances are it's a mutant—called a "sport" by plant breeders

—that may have some new qualities that are of commercial value. For most sports, it doesn't work out that way, yet some of today's most popular fruits—such as navel oranges—did arise as chance mutants. What this amounts to, of course, is one way of taking advantage of Mother Nature's chance changes, using a natural, haphazard method of genetic roulette.

What should be reemphasized, however, is that most mutations turn out to be losers, since they usually occur in what are already well-run, thoroughly organized, balanced systems. Mutations usually make the individual less able to cope with the environment he finds himself in. Improvements are the exception, but they are an important exception that has been responsible for progress in the different species.

This, then, is how evolution has proceeded over billions of years, slowly selecting out good mutants that help strengthen a species, killing off the detrimental mutants which can't compete. Such an approach to progress may seem wasteful both in terms of time and materials, but then nature has never been known—in the long view—for being stingy with either.

What is it, then, that causes mutations?

Biologists are now convinced there are many causes, and some of these more powerful causes of mutation have been well demonstrated. Of great importance, of course, is radiation. Indeed, the earth is constantly being bombarded by invisible energetic particles coming in from space. Some of the weaker particles come directly from the sun, but other immensely powerful particles arrive from deep in space and are usually referred to as cosmic rays. In addition, it is known that most creatures on earth also receive significant doses of radiation from naturally radioactive substances like uranium and thorium in the rocks and soil.

In any event, some of these speeding subatomic particles will pass through living tissues and hit chemicals in the chromosomes, causing changes in the chemistry of the nucleic molecules by altering the bases in DNA. This appears to be a continuing, even unavoidable process, but it is also known that living organisms have become used to it and have evolved systems for coping with most radiation. In relation to the effects of radiation on life, one of the current ideas about aging is that the repair mechanisms that correct radiation damage eventually slow down and are not able to keep up with the rate of change. This would allow more of those deleterious changes to accumulate, eventually leading to loss of control and the death of the individual.

A large number of chemicals are also known to cause mutations. Indeed, some chemical compounds used in modern foods—such as one important red dye—are being found to be mutagens, capable of causing mutations, and are thus gradually being removed from the market because of the suspected danger. This, of course, is one of the important functions of the U.S. Food and Drug Administration, and of similar agencies abroad. One of their main tasks is identifying and controlling substances added to foods, marketing of drugs and sales of cosmetics which might be dangerous to the body, or harm the genes. A sad case in point, certainly, was the drug Thalidomide, originally sold outside the United States as a tranquilizer. When taken by pregnant women—to help cope with discomforts accompanying the early weeks of pregnancy—Thalidomide seems to have interfered disastrously with the genetic program that was building up the new babies, disrupting what the genes did as the cells multiplied into the billions. The result was a large number of children born with deformed limbs; arms, for instance, that resemble a seal's flippers. Fortunately, for Americans at least, the Food and Drug Administration had not permitted Thalidomide to be marketed in the United States.

Returning to the causes of mutations, both radiation and chemicals, despite their talents for causing problems, have served as very powerful tools in the continuing study of how genes work and how they control the construction and operation of a living creature. One chemical, colchicine, can briefly stop the process of cell division in plant tissue without killing the whole plant. What colchicine also does, however, is scramble the genetic controls. When colchicine-dosed cells resume dividing again, some of the new daughter cells end up with double the normal number of chromosomes. By this method, then, plant experimenters have found a way to produce vigorous new types of plants such as snapdragons that far outperform normal snapdragons containing the normal—or diploid—number of chromosomes. Plants that have had their chromosome number doubled and redoubled, then, are known as polyploid plants.

It should also be mentioned that chemicals are being used almost routinely now for inducing mutations in creatures such as the fruit fly, in root-invading worms called nematodes, and viruses. In the studies using fruit flies and nematodes, scientists have been searching for mutants that behave differently than their relatives. Scientists are trying to resolve that old question over what role the genes play in the control of behavior. This is known as the argument of nature vs. nurture.

But this represents only one approach. Scientists have also done more than induce just a single mutation. They've progressed to the point where they can tie strange behavioral mutations to more obvious characteristics like a fruit fly's bristly hair, stubby wings, or an off-color body. Thus they can spot, in a huge group of fruit flies, those that carry the behavioral mutation because it is linked to the mutation for this other, visible characteristic. They've also been building what are called "mosaic" animals; fruit flies, for instance, which are born half-male, half-female, with the dividing line running right down the midline of the body. They've also succeeded in building flies that are born with two sets of circadian rhythm, the daily rhythm of sleep and wakefulness. In some cases the head was set at one speed, the body tending toward the other.

Another approach to studying the genes was used by a group at the Salk Institute in La Jolla, California, headed by Dr. Renato Dulbecco. Dulbecco, who has since moved to England, worked with his colleagues at Salk to cause mutations in what is known as a DNA virus, a virus that goes by the name SV40 (for simian, or monkey, virus 40). This virus is known for its ability to turn normal cultured cells—grown in a petri dish—into cancer cells. The virus first invades the cell and then mixes its own DNA into the cell's DNA. Thus the virus's DNA assumes command of the cell's genetic material, reprogramming it to obey the virus's new set of instructions. In some instances, too, the virus makes the cell become malignant, and the victimized cell begins dividing rapidly, out of control.

Dulbecco and his colleagues produced mutations in a great number of viruses. They selected a mutant that was sensitive to temperature and found that above a certain temperature the virus invaded the cells and transformed them into cancer cells. Below that temperature, however, the virus was apparently dormant, and the cells it had invaded and transformed turned back to normal cells. Thus what they had devised was a useful "switch" by which they could simply turn the characteristics of cancer on and off in a cell, merely by changing the temperature. This technique, of course, is now being widely applied in other living systems— although primarily still in viruses—because of the great opportunity it offers for learning how cancer can be induced.

Radiation is being used both for causing mutations and for destroying the breeding ability of some insects. Experiments aimed at producing beneficial mutations in food crops—essentially by bombarding everything in sight with X rays or some other form of radiation—have been done all around the world, but have not met with much success. Experimenters have been able to produce visible radiation damage, but the number of beneficial mutations produced has been disappointing.

Using a different approach, scientists have found better success in use of radiation for sterilizing adult insects. In this method, a great batch of males, such as screwworm males, can be irradiated, then are released in the field to mate—unsuccessfully—with wild females. By saturating an area with sterile males, most of the normal wild males are displaced. The result of mating with the sterile males is, obviously, a smaller population of new screwworm flies. This technique at first proved to be successful, and was seeing widespread use against many other pests which annually destroy billions of dollars worth of commercial crops, but later it was found that laboratory males were less competitive, and screwworm populations rebounded.

It should be noted that these approaches to genetic manipulation represent rather crude scattergun attacks. In the future, genetic engineering is expected to be much more precise and specific, replacing particular genes in individual plants, animals and people. Already coming up are new techniques in which single genes can be handled relatively simply and can be inserted into the genetic machinery of almost any creature. What this involves, however, is the difficult task of finding and then manipulating one or two specific kinds of molecules out of a soup made up of millions.

Of course, this is where the revolution has been. Powerful new methods and chemical tools are now helping scientists locate specific genes, pull them out of living tissues, and then plug the genes back into living bacteria. The bacteria are being used as "gene farms," allowing scientists to grow as many copies of the gene as they want. All they need to do is supply food and allow the bacteria to multiply normally. So, for the first time, biologists now face the prospect of having enough genetic material available to do some important experiments, to let them look at specific genes in fine detail.

This development could—and probably does—provide one of the crucial steps that will carry the biological sciences toward the goal of true genetic engineering. Indeed, what biologists would hope to do by using such techniques on living creatures might most resemble what a California hot-rod mechanic does with an ordinary automobile engine. Even if given a car that runs perfectly well, smoothly and reliably, the enthusiastic hop-up artist still feels compelled to somehow make it better, or at least different. Changes he would make range from simple adjustments like fine tune-up work to more drastic approaches like boosting the engine's compression, changing the valve timing, or even enlarging the size of the whole engine.

Similarly, biologists are looking forward to doing significant tinkering on living machines, but they won't be working with camshafts, carburetors, and distributors. Their targets will be organisms that are alive and breathing, and their tools will include complex chemicals, strange enzymes, and new combinations of genes inside living cells. It seems obvious, however, that they will end up building some strange new hot rods, high-performance creatures never before seen in nature.

The goal, of course, is to venture several steps beyond anything Mother Nature normally attempts or even allows. Scientists are already putting strange genes into creatures where they'd never occur naturally, and this is just the beginning. Scientists will continue battering down one of nature's most ancient, rigid barriers, the barrier that blocks crossbreeding between different species. You can bet, indeed, that biology's hot rods are going to be something spectacular, and something to contend with.

Strong hints that this kind of thing might someday be possible —that the genes themselves might indeed be manipulable—arose in the early 1950s when it was discovered that the strange substance called deoxyribonucleic acid (DNA) is the basic chemical responsible for carrying hereditary information. According to early discoveries, and to the Nobel Prize–winning work of Sir Francis Crick and James Watson in England, DNA is composed of those nitrogenous bases which, when combined to resemble a ladder, form a long molecule which twists and coils around itself to form a tight helical structure now famous as the double helix. So now, after years of slow, painful, often frustrating and disappointing work, bioscientists have finally reached the point where they've started deciphering how the DNA molecule works inside the living cell, how it's made and in some ways how it's controlled.

Much more recently Dr. Herbert Boyer and his coworkers at the University of California School of Medicine in San Francisco have isolated a special enzyme which, under the right conditions, can be used to snip bits of DNA into small segments. Such segments may vary in length, but each is known to carry anywhere between one and ten complete genes. The important point— discovered by Janet Mertz and Ronald Davis at Stanford University—is that these tiny DNA fragments are not snipped cleanly, smoothly, but that they have little single-strand tails left that are now referred to as "sticky ends."

Sticky ends, indeed, have turned out to be of utmost importance for genetic engineering research. Clever biochemists now use them as connectors for tying foreign bits of DNA to other pieces of DNA, to other genes. This would resemble the way railroad cars are assembled into a long line to make up a whole train. Later, too, other special enzymes were found that can be used for reconnecting these bits of DNA, recombining the molecules that had been snipped apart.

Again, it should be emphasized that scientists are now successfully doing what Mother Nature has previously forbidden: mixing the genes of different organisms, mixing genetic information from creatures as different as sea urchins and yeast, as widely divergent as bacteria and man. For the first time, indeed, genes from such disparate organisms are finding themselves tacked together on the same chromosome, doing what comes unnaturally.

This may end up sounding slightly preposterous, but for the first time some of those silly old jokes, which go like: "What do you get if you cross a goat with an owl? A hoot-nanny!" are on the verge of being possible if someone wants to try. Biologists are approaching the point where they may supply some of those silly answers, but the results may not turn out to be all that laughable.

While the discovery of those talented snipping and sewing enzymes was important, of equal value was the gradual discovery that such new strings of genes can actually be plugged into living organisms to be tested, to see if the genes put in can be expressed. The really important step came when it was recalled that bacteria —which don't keep their chromosomes locked up inside a tiny nucleus—also contain small doughnut-shaped rings of DNA called plasmids. These plasmids are found wandering around inside the bacteria, but they are distinct from the bacterium's main source of genetic information, the chromosome. The plasmid rings were discovered years ago, but were thought to be of small significance and little interest, until recently.

What was discovered, of course, is that plasmids—which are merely small, naked rings of DNA—can be pulled out of bacteria, snipped into smaller chunks, then reassembled. What's more, it was found that small pieces of foreign DNA can be neatly inserted, without obvious damage to the genetic codes, into the plasmid ring. Better yet, when the new, recombined DNA ring— the new hybrid plasmid—is pushed back into a bacterium, it appears to be tolerated as part of the creature's normal complement of DNA, dividing and multiplying right along with the whole organism. In addition, there is also some evidence that even samples of foreign DNA can be "read" by the bacterium's gene-reading system. But there are also hints that foreign genes don't work well inside primitive creatures where they don't belong.

Nonetheless, even if newly inserted genes from higher animals never do work properly inside a bacterium, this technique for adding new genes where they don't belong will still be useful, because experimenters will still be able to get bacteria to divide and redivide, building large new supplies of a particular gene that would otherwise be difficult to isolate in quantity. Thus, despite its inability to act on genetic instructions, a bacterium will still produce abundant new copies of the inserted gene just as it produces copies of its own genes during division.

This may not sound terribly exciting to the layman, but for scientists who've been waiting years to get their hands on large amounts of a single gene's DNA, it represents an important achievement in overcoming one of their major problems. Now, for the first time, they can begin studying single genes in great detail to find out how the genes are put together, and how different kinds of genes compare in detail.

How individual genes differ structurally is certainly one of the important areas scientists are hoping to explore because elegant work done already on easier-to-obtain molecules, such as enzymes, has shown how important and interesting this work can be.

A group of chemists at the California Institute of Technology, for example, has used an important metabolic chemical called cytochrome-C to look deeply into the involved, complicated process of evolution.

What they've done, as have others studying different chemicals, is to look closely at the sequence in which amino acids are arranged in the long chains that make up proteins. Amino acids, which are the building blocks of proteins, are normally strung together in long chains to make up protein molecules. The order in which they are strung—like different-color beads on a necklace —is important, since the sequence they are strung in governs how the long protein chain will fold up. This, in turn, governs the shape, size, and the biological activity of the protein molecule. Counting the order in which amino acids fall in a string of protein, then, is known as "sequencing" the molecule. It is a process that is coming into more and more widespread use as scientists increasingly recognize its value.

An important side benefit of this sequencing work, other than revealing how the molecule itself is built, is to provide an interesting look into what has been happening over the ages in the process of evolution as ever more complex organisms evolve from primitive creatures.

For any given biological molecule—such as cytochrome-C, the hemoglobin in blood or the proteins called histones found inside the nucleus—the sequence, or order in which amino acids are strung, tends to be very stable because, again, the sequence governs how the molecule folds up. Folding must be precise and stable, or the molecule may not be able to perform its function in the living body. What this says is that these important molecules can't stand much alteration, much change, much mutation, in the genes that build them, or they won't work. Mutations, nonetheless, do occur over the ages. Creatures that suffer too-drastic mutations don't survive, but some less-severe mutations can occasionally be allowed in less important spots on the molecule. When comparing two animals, the number of such small changes that are found on a particular kind of molecule—on a molecule like cytochrome-C, which is playing the same role in both creatures—seems to reflect both the amount of time since two creatures shared the same ancestor, and the distance they are from each other in evolutionary terms.

For instance, hemoglobin must fold up properly so it can accept an oxygen molecule in the lungs, hold it while traveling in the bloodstream, then release it at just the right time to help keep tissues alive. Similarly, cytochrome-C, a molecule that's important as an energy carrier in the process of metabolism, folds up so that one side has a deep crevice, rather like the pocket in a first baseman's glove. This crevice must be able to open up slightly at just the right time, then close again to carry an electron from one reaction to another. In all species where it is found, then, cytochrome-C performs this same function, in the same way.

Based on these findings, then, biologists and biochemists have determined that a particular molecule like cytochrome-C—which performs the same task in animals as distantly related as the horse and a skipjack tuna fish—must still have basic similarities in amino acid sequence. Yet, despite these important similarities, the differences such molecules from different species do show is one way of measuring how far apart they are in evolution, how far back in the past it was that they shared a common ancestor. What this says is that the evolutionary distance between the fish and horse must be very great not only because of their physical appearance, but also because of the large number of amino acid changes seen at specific locations. The same evidence says, too, on the other hand, that the evolutionary distance between the horse and mouse is much less, since the changes in amino acid sequence are fewer, and since both are mammals. When it comes to assessing the distance between man and chimpanzee, there's hardly any difference at all, if a difference of one or two amino acids can be found. What this implies is that man and ape separated from a common ancestor only a short time ago—at least in terms of evolution—and that in biochemical terms they are almost the same animal.

This process of studying evolutionary distances can be taken even farther with universal molecules like the histones, which are specific types of proteins found associated with the chromosomes in a cell's nucleus. By sequencing histones, looking at the arrangement of their amino acids, it is even possible to determine how closely—or really how distantly—species as diverse as yeast cells and cows might be related. The distance, of course, is enormous, as told by the shape, size and activity of the two organisms, and by the number of changes in amino acids. Nonetheless, it does say that way, way back in earth's history, both the cow and the yeast arose from a common ancestor.

What this whole thing amounts to, then, is a sophisticated biological yardstick—or a very durable stopwatch—that tells us how much time has passed during evolution of the species, and how long ago in the past we shared common ancestors.

Unfortunately, such a yardstick tends to assume that the rate of mutation has been relatively steady for the creatures on earth, and that the rate has been similar for all species. These two assumptions may not be valid, so the accuracy of that biological stopwatch could be open to serious question.

How, then, does all of this relate to the problems and promise of genetic engineering?

It's simple. Biologists would like to perform the same sort of chemical tricks on the genes, sequencing the chemicals that are found in specific portions of the genetic code. They would hope to compare the genes that do the same things in different species. One important goal, certainly, would be to analyze the basic chemical sequences of genes that are found inside cancer cells, then compare them with the chemical sequences in the genes taken from normal cells. It should be interesting—and perhaps important—to see what the differences will be. Will scientists find that one simple nucleotide being changed in only one gene can account for the disaster known as malignancy? Which gene—or genes—would be responsible? Is cancer caused by the actual mutation of a gene, or is the inclusion of a piece of foreign DNA—perhaps from a virus—the cause?

Such questions seem almost endless, but at long last, it appears that scientists have found powerful new methods for finding some of the answers—or at least for asking some important new questions.

One note of caution is necessary, however. As in all of science in the past, exciting new advances in biology will almost certainly end up creating more questions than they answer. Still, it seems possible now that some of the older, perhaps more fundamental questions about life, about sickness, will finally yield some answers. It is no wonder, then, that scientists are becoming excited, and no wonder that some of these important new experiments are already under way.

One of the factors which allows this work to progress very rapidly is that bacteria, such as *E. coli*, multiply very rapidly. *E. coli*, for example, divides about every twenty minutes, so that a scientist trying to grow up a specific set of genes by inserting them into the bacterium can now have billions of copies within a few days; copies that are available in relatively pure form and in massive numbers.

A second important approach to the same goal involves the use of viruses. This requires taking the DNA from the virus, cutting its doughnut-shaped ring, then combining it with the special bit of DNA that is of interest. This is all done using much the same chemistry that was developed for working with plasmid rings, but it differs in the fact that the virus, carrying a new piece of DNA, is sent to attack the bacterium and thus inserts its DNA into the bacterium. Once inside, the virus's foreign DNA molecule—and the additional piece that the scientists slipped in—are faithfully copied and reproduced along with the bacterium's own set of genetic blueprints. Again, this also amounts to one way of stuffing a particular piece of DNA into a living, dividing bacterial cell, where it can be easily reproduced in large quantities.

Another important point that should be mentioned is the advantage scientists see in using a particular organism like E. coli. Currently, E. coli is the best-understood creature ever to grow up in the laboratory. It is also a very simple organism, and is found normally living happily in the human throat and intestine. It can also be grown on a simple diet of sugar, nitrogen and a few minerals. Lastly, E. coli is relatively easy to handle, and it is not normally considered dangerous.

Using E. coli as a microscopic-size brood mare for reproducing odd bits of DNA is just one early step in this work, however, and according to Dr. Paul Berg of Stanford University, this type of genetic transfer—the insertion of new genetic materials into an unrelated organism—will probably also be done later in the cells of higher organisms, using the genes from any other kind of organism. It is possible, Berg believes, that plasmidlike DNA rings will also eventually be found occurring naturally inside mammalian cells. If so, these, too, will be used for carrying new information, new DNA, into the cells of higher organisms such as man. Even if such plasmidlike bodies don't exist in mammals, however, the promise is still strong that viruses—which were well designed by Mother Nature to carry genetic information into the innards of cells—will be used to ferry new genes into organisms where they don't naturally belong.

Despite the new advances that are already in hand, though, one

Dr. Har Gobind Khorana, at the Massachusetts Institute of Technology, worked with his colleagues for almost a decade to build the first functioning artificial gene. Once the gene was found to work properly inside E. coli bacteria, plans called for doing experiments, altering small portions of the gene, to see how precise changes alter the action of the gene in the living bacteria.

of the persistent problems has been to find ways for identifying, isolating, and purifying the tiny genes that scientists would like to work with. While the task still isn't simple yet, important steps have been taken toward this goal, already making identification and isolation simpler than before.

At the Massachusetts Institute of Technology, Dr. Har Gobind Khorana has already succeeded in the slow, painstaking task of building an active gene artificially. It is a copy of a gene found in *E. coli*, a gene normally responsible for building one type of transfer-RNA. Khorana, who received the 1968 Nobel Prize in physiology and medicine for his biochemical research, led a team of scientists in synthesizing, piece by piece, a 126-unit gene, including the chemical "start" and "stop" signals needed to make it function. It was the first gene ever synthesized, and it was found to function detectably inside a living bacterial cell.

Completion of this difficult task—the creation of a working gene, complete with "start" and "stop" signals—was announced at a meeting of the American Chemical Society, held in San Francisco, in 1976. The announcement culminated nine years of work that Khorana once characterized as "powerful, but tedious." Six years earlier Khorana and his colleagues had announced the synthesis of the first artificial gene, a 77-unit yeast gene that is also responsible for building one type of transfer-RNA inside a living cell.

The M.I.T. team began building their *E. coli* gene by hooking together small ten- to fifteen-unit segments of DNA that were made up of the individual nucleotide building blocks. They then stitched these together to form four larger segments of the gene, which they finally linked together to form the whole structure.

Obviously, this same technique might be extended for use with mammalian genes—even human genes—except that instead of being only 126 nucleotides long, the human gene, typically, is millions of units long. Thus it would become a difficult, almost unending task to build up even a single human gene. Despite this problem, it can be said that Khorana's work has provided strong proof that working genes can indeed be built up artificially.

In the meantime, however, other teams at other institutions have been pushing their own research, and already another ap-

Synthetic Tyrosine tRNA Gene

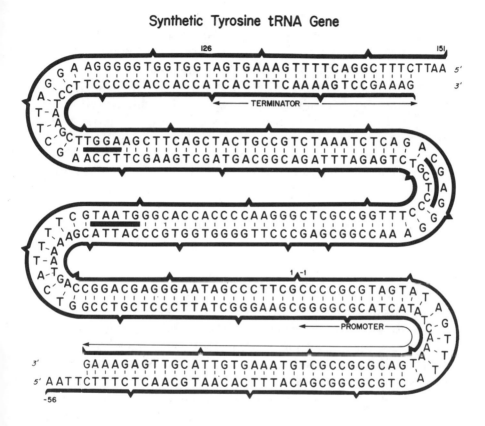

The diagram represents the structure of the first artificial gene, as made by Dr. Har Gobind Khorana and his co-workers. It was also the first artificial gene to work inside living E. coli bacteria. This diagram shows the gene's control elements—the promoter and terminator—which tell the cell's protein-building machinery where to "start" and "stop" reading. Segments between the points in the diagram were first synthesized chemically, then were joined to form the complete DNA molecule, the double helix.

proach to building genes artificially has been developed by scientists at Harvard University. Moreover, this team, using a different method, was able to artificially build the first synthetic mammalian gene. They did the job backward by using important enzymes stolen from a special kind of virus. The important point is that this novel approach has opened the door to the possibility that any gene from any of the higher animals can soon be located, copied, and grown in quantity for study in detail.

According to members of the Harvard team, the ultimate goal is to understand how genes are expressed in higher creatures. A step in that direction is provided by their new technique which gives scientists one potentially important way to isolate specific genes for study. They explained: "The chromosomes of higher animals contain thousands of genes, most of which never function. To try to purify any one particular gene from this multitude is like finding a needle in a haystack. Yet, unless we have a substantial sample of a gene, it is very hard to conduct proper research into it."

The gene they chose to copy and synthesize is responsible for production of hemoglobin in rabbits. Hemoglobin is a key ingredient in red blood cells, which are responsible for carrying oxygen to all parts of the body.

Under normal circumstances, the cells that produce new red blood cells "read" a section of their DNA code which includes the instructions for making the hemoglobin molecule. In this reading process, the enzymes go to work building a single-strand RNA copy of the gene, a one-legged copy of the ladderlike DNA molecule. Then the new single-strand RNA molecule, bearing the chemical message transcribed from the DNA codebook, moves off to help assemble what is referred to as a protein-making factory— a polysome—around itself. The polysome then gets busy stringing the proper amino acids together to build a protein molecule, doing this as instructed by the messages that were read onto the RNA molecule, the messenger.

This process illustrates the normal direction of flow of biological information: the chemical information required for building and controlling the living organism goes from the original codebook—

the DNA—to the messenger molecule—the RNA—and then to the factory—the polysome. The product of this chain of events—in this case, the hemoglobin in a rabbit—is then released to do its work inside the living body.

Biologists have learned, however, that there are some important exceptions to this basic rule; that biological information flows from DNA to RNA to protein. They have known for years, in fact, that some viruses—and especially a few that cause cancer in birds and mammals—carry no DNA of their own. They carry *no* double-stranded genetic codebook of their own at all. Instead, these viruses, known as the C-type viruses, merely carry a simple single-strand bit of RNA that contains all of their own virus-building instructions. This discovery, like numerous other discoveries in science, ended up raising more questions than it answered, the main one being: how can a virus invade a cell and then, using just this simple strip of RNA, take command of the cell and subvert its whole DNA system? The question was eventually answered by Dr. Howard Temin of the University of Wisconsin, and Dr. David Baltimore at the Massachusetts Institute of Technology. They were jointly awarded the 1975 Nobel Prize in physiology and medicine for their work.

They discovered that along with their RNA-based set of genetic blueprints, these special C-type viruses also carry an enzyme that can make the genetic information flow backward. This enzyme, known as reverse-transcriptase (or, tongue-in-cheek, Balti-Teminase) makes the invaded host cell use the virus's RNA molecule as a template for building a new DNA molecule, which is then inserted into the host cell's chromosomes. By this method the virus takes control of the cell and what it builds—which, of course, is more and more of the virus particles. This, one might say, represents a lazy, sneaky and very efficient way for the virus to reproduce. It merely subverts the host cell's machinery, causing it to build more viruses which, in turn, go off to infect other cells in the same manner.

Getting back to the problem of making synthetic genes, the Harvard group borrowed this backward-running information system that the C-type viruses use for reproduction. The enzyme was carefully put to work by the scientists for building their own artificial DNA molecule, the rabbit's hemoglobin gene. And now, with the technique essentially all worked out, members of the Harvard group think it should become possible to build artificial copies of almost any gene they're interested in.

In constructing their artificial rabbit gene, the Harvard scientists first searched around for a natural RNA molecule from the rabbit's hemoglobin gene to use as a template. Then they mixed in the virus's enzyme, the reverse-transcriptase. Into the pot, too, were tossed the chemical building blocks for making a new DNA molecule, the individual nucleotides that would be strung together to make up the new ladderlike DNA molecule. By this process, the Harvard scientists demonstrated for the first time that artificial synthesis of a DNA copy of a complete RNA molecule is possible in the test tube. In essence, what they had done was to perform the same trick the C-type virus performs when it invades a living cell. The scientists, however, did it in a test tube.

Despite their success, though, the experiment wasn't complete. The Harvard group still faced a problem that kept them from synthesizing the complete gene. After their first success with reverse-transcriptase, they found that they had, indeed, constructed a DNA molecule, but it was a faulty molecule that had only one strand rather than two, like a one-sided ladder, or a train track with only one rail. They also found that their synthesized molecule had a strange hook—or small half-loop—attached to one end.

Solving this set of problems took another nine months, but the team finally found another enzyme, called DNA polymerase I, which could locate that odd hook in the solution and use it as the starting point for building the second strand—the second side of the ladder on the DNA molecule. Next, a third enzyme was used to trim that excess loop off the end of the gene.

The importance of such a development to the whole budding science of genetic engineering would be hard to overemphasize. This step implies that scientists can now begin seeking out any RNA molecules—such as those involved in making insulin, or the cells that build human hemoglobin, or human growth hormone—and use these RNA strips as templates for building the genes that code for these important products.

The next step would be to combine the two technologies, to shove these new artificial genes into living cells—beginning probably with a bacterial cell like *E. coli*—and find ways to grow up vast numbers of a specific gene. Better yet, if they can induce *E. coli*'s cellular machinery to "read" these foreign genes, their next step would be to use the bacterium as a tiny factory to produce insulin or growth hormone. Even if this doesn't work, scientists would still gather enough copies of specific genes to begin sequencing them to see how they're put together.

An important and perhaps dangerous step, too, would be to grow up enough copies of cancer genes and normal genes for study, so biologists can finally begin learning what the difference is. This may not lead to a quick solution to the problem of cancer, but it should give medical men a strong new lead toward what causes cancer.

These, then, are some of the steps that are already being taken, and they will soon inevitably lead to more developments of even greater significance. Some biologists suggest it won't be long now before the first man-made mouse is built, similar to John Gurdon's cloned frog.

It should be clear, too, that the world of the genes is now gradually, even painfully, opening up to the men of science. What they'll eventually find—and what they'll eventually do with the knowledge—can't be predicted. It will almost certainly be exciting.

What's more, it is going to be a challenge. All societies are now faced with the problem of how to deal with a science that has the power to drastically alter the living things of this world. But is the world ready for this? Can we use these developments wisely in medicine, in agriculture and industry without destroying each other? Can we devise new species of bugs, birds, mammals—even men—without ruining the quality of life as we know it?

Perhaps more important, should we be spending our time and

resources on such esoteric science when most of the world is just plain hungry, when we're not even able to *apply* much of the science we already know toward solving this world's problems?

Unfortunately, these kinds of questions won't be answered in the laboratory. But the work on genes is going to continue, and it probably means we've already taken the first tentative steps into Aldous Huxley's Brave New World.

3

Medicine: Expect a Miracle

TIMIDLY, QUIETLY, the young couple stepped in, settled into soft leather chairs, then stared across the bare brown expanse of the doctor's desk. Obviously, both were prepared for the worst.

"And what do the tests show?" the woman asked quietly, almost inaudibly. "Can we have our baby?"

The answer came quickly as a grin from the doctor who, leafing through an open folder, picked out a few charts and a typewritten sheet.

"It seems you've done all right," he finally answered. "Your baby will be normal, but we do find hints she'll be a Tay-Sachs carrier."

"She?"

"Yes, it's a girl. In tissue studies, we found the cells have two X chromosomes, so your baby is going to be a girl."

"But what about Tay-Sachs disease?"

"You were very lucky this time," the physician answered. "Since both of you are carriers of the Tay-Sachs gene, as I explained before, you had a twenty-five percent chance—one in four—of having a baby with Tay-Sachs disease. These odds will, of course, be the same for your next child if you decide to have another."

If his answer had been different, if the genetic lottery had dealt their child two genes coding for Tay-Sachs disease, their child, if allowed to be born, would have faced severe mental retardation, blindness, and early death. For now, this disease—which is found most commonly among the Ashkenazic Jews—is wholly incurable.

These parents, in that event, would have faced their own new dilemma. They would have been asked to decide, rather quickly, whether to let the pregnancy progress to full term and delivery, or terminate it immediately with an abortion.

While their answer is obviously important to them and their unborn child as a family, in terms of the science of genetic engineering, the important thing is that they have that choice at all. Now, for the first time in history, couples who have good reason to suspect their unborn child carries an important genetic defect can—in many instances—find out well in advance if their suspicions are correct; find out in time to do something about it. What they choose to do about it, of course, is often a terribly difficult, painful decision, but it is at least possible now to give them that choice.

The Tay-Sachs couple knew from both of their own family histories that each might carry copies of that hidden defective gene. Indeed, both had been tested and were carefully warned of the risk of having children. They had chosen to take that chance and have children of their own despite the genetic odds. They could instead have chosen to forego having children of their own and adopt instead.

Once the woman became pregnant, however, their fears began multiplying. That one basic question: "Is the baby going to be all right?" kept nagging them.

Fortunately, again for the first time in history, there is now a way to diagnose the unborn baby for genetic diseases. In a recently developed procedure called amniocentesis, doctors take a sample of the amniotic fluid surrounding the baby, collect loose cells they find in the fluid, grow some of these cells up into tissue culture, then test for the disease. Amniocentesis, however, is still considered by some a bit dangerous, since sampling the amniotic fluid may slightly increase the risk of spontaneous abortion. Nonetheless, it is a procedure through which doctors can at last forecast that certain genetic illnesses will or will not strike.

This, then, is one of the important developments coming from laboratory work which might, in one sense, be called genetic engineering. Such test-tube work is continuing, and according to biologists, there's still much more that will come out of the laboratory that will be applied directly to medical practice.

As one would expect from scientists, however, most of the work now being done under the flag of genetic engineering is truly basic biology or biochemistry, science being done for the sake of science, for the sake of unraveling little bits of Mother Nature's huge, intriguing riddle called life.

Scientists, indeed, are fond of the word "elucidate," which means to make clear, to understand, to explain. This is what they're doing in laboratories as they probe and tinker with the genetic materials taken from the cells of various living creatures. They're hoping to learn ever more about how these living things are built, how they're maintained, and how they're controlled.

One thing should be obvious: the sciences involving genetic manipulation aren't going to stop just at the point of mere understanding. As with nuclear physics—where all the action certainly didn't stop at the point of understanding, but went on to weaponry and nuclear power—genetic engineering will also be developed so it can be used, perhaps even routinely. Once given the tools and the ability to actually perform genetic engineering in humans, doctors who so often see the sad consequences of genetic diseases will begin—perhaps out of desperation—trying to apply more of the lessons of molecular biology to living patients.

It's evident, too, that when it comes to applying these new techniques to humans, the possibilities are enormous. Some visionaries are already looking far ahead toward building new, better kinds of humans, whether they can define what "better" means or not. Some of these ideas, which may seem a bit screwball at first glance, include building a race of smaller, more efficient human beings, while others would build in superbright minds or other extravaluable abilities. Such uses of genetic engineering, however, seem very distant, even if desirable. Certainly there are things that can and will be done sooner, and some things will be done almost immediately if one form of genetic engineering—described better, perhaps, as cellular engineering—is encouraged and developed.

As should be expected, most of the hopeful, compelling speculation about what genetic engineering may do revolves around the practice of medicine—rather than rebuilding the species—and what physicians hope genetic engineering will be able to accomplish against a background of some 2,000 known genetic diseases.

Such diseases are well known. They range from relatively common but nonlethal birth defects, such as cleft palate and clubbed feet, to the more disastrous maladies of mongolism, phenylketonuria, Tay-Sachs disease, sickle-cell anemia, hemophilia, and, perhaps, even diabetes. Indeed, records indicate that some 250,000 babies born annually in America show some sort of genetically based abnormality.

Even without genetic engineering, of course, modern medicine has made some impressive gains against a few of these inborn diseases, especially against hemophilia, diabetes, and phenylketonuria (PKU). The medical advances made so far, however, merely amount to management of these conditions, not cures. No real cures are yet in sight for such diseases without some resort to genetic engineering—or at least cellular engineering.

Still, all these hopes, efforts and suggestions about combating genetic diseases directly—by tinkering with a person's genes or tissues—represent only one aspect of what modern biology will be doing in the field of medicine. These ideas say nothing about how genes implanted in different kinds of microbes will be put to work making cheaper, better new medicines, hormones, and other chemical substances that will have direct application for people stricken with disease. The impact from this one area alone should equal or exceed anything that will be done for patients treated through genetic manipulation techniques.

And, finally, it should be realized that a good, basic understanding of life's most intimate processes will eventually give scientists and doctors more vital leads toward the causes of cancer, heart disease, and emphysema.

There are several classes of genetic diseases. The great majority of these genetic disorders, caused by malfunctions or misreadings of the genetic instructions, turn out to be so severe, so catastrophic to the developing child, that the infant never even gets born. This accounts, in part at least, for the fact that 15 to 20 percent of pregnancies in America end in spontaneous abortions, miscarriages. Thus Mother Nature steps in much of the time to do her own abortions well before life ever really gets under way. The fetus apparently finds its genetic blueprints so severely scrambled, so much in disarray, that life itself is impossible beyond certain stages of development.

There are, however, those few babies who do get born despite their scrambled genes. Thus we do see living mongoloid children, tiny infants with Tay-Sachs disease, others with PKU and males with hemophilia. The classification of these and other genetic diseases follows rather simple logic, since they are divided mainly according to how soon they kill the victim. First, as mentioned, there are those very severe cases which never get born. Next are the genetic diseases which kill the infant soon after birth, such as galactosemia—a defect in the baby's ability to metabolize the milk sugar galactose. This condition usually leads to death within two weeks of birth, since a newborn child's diet is usually all milk. Then there are illnesses like Tay-Sachs disease, leading to severe mental retardation, blindness and, invariably, death by about age four. Tay-Sachs, since it is most commonly found among Ashkenazic Jews, is one of those diseases—along with sickle-cell anemia among blacks—which appears to be largely confined to well-defined population groups. Such diseases lend themselves rather well to screening programs aimed at detecting carriers of the defective gene.

The next category of genetic diseases, such as thalassemia—also called Cooley's anemia or Mediterranean anemia—usually let the victim live longer, most often to about age twelve. Thalassemia causes abnormally thin, or low, red blood cell count. A child can be kept alive longer through the use of blood transfusions, but never really has a chance to lead a full, normal life.

Still other genetic diseases hold off until a person reaches adulthood before developing, such as Huntington's chorea. This disease, which strikes both men and women, involves degeneration—indeed, disintegration—of the brain. Symptoms usually strike after age thirty or forty, but in some rarer cases Huntington's chorea is also seen in children or the aged. As the disease progresses, the victim shows changes in personality, becomes moody and obstinate. Ability to walk deteriorates, and death normally follows onset by about fifteen years. No treatment is known, and the mental processes become so disturbed that suicides are common among Huntington's chorea patients.

Added to these obvious genetic diseases, of course, are the more common, more subtle maladies like degeneration of the heart and blood vessels, the lung diseases like emphysema, and cancer. All of these appear to have some still undefined genetic factors, and they typically wait until their victim is well into adulthood before striking. Onset of these ailments, too, may also be linked with diet, with environment, or occupational hazards.

Even though these disastrous genetic diseases like hemophilia and galactosemia are considered incurable, they're not all untreatable; their deleterious effects can be avoided to some degree. In recent years, scientists and doctors have made important strides toward at least minimizing some of the results of a few genetic diseases. If diagnosis is made in time, especially in cases like PKU and galactosemia, much of the damage to the newborn child can be avoided simply by eliminating certain foods from the diet.

Perhaps the best example of intervention for the control of genetic disorders is found in the treatment of diabetes. Doctors really aren't convinced that diabetes is actually a genetic disease, but it is a disease in which the body, for some reason, doesn't get enough insulin or, if there is sufficient insulin, the cells aren't reacting to it the way they should. Two doctors—Frederick Banting and Charles Best—discovered back in the 1920s that diabetes can be managed successfully by injecting additional insulin at regular intervals, and by keeping tight rein on the diet. Now, while there is still much disagreement over whether diabetes is a true genetic disorder, there is no quarrel whatever about insulin treatment enabling thousands of people to live near-normal lives.

As a disease, diabetes may even represent a number of similar disorders or maladjustments of the body's delicate chemical system that is responsible for metabolizing sugars. When not enough insulin is present, sugars accumulate in the bloodstream causing symptoms that may include extreme thirst, hunger, weakness, and weight loss.

Normally, for a person whose blood sugar is in proper balance, tiny cells known as the beta cells, growing in small glands called the islets on the pancreas, release insulin for the body's needs. This occurs in response to signals that sugar in the bloodstream is too high, in response to warnings that the body needs to burn up more sugar. When this system is out of balance—with either the beta cells not producing enough insulin, making incomplete or unusable insulin, or the body's cells not responding properly to insulin—diabetes is the result. The response from the medical profession, then, is to supply additional insulin in order to bring sugar metabolism back under control. And the best part is, it works.

Despite this impressive success, however, many diabetics find themselves riding a hormonal roller coaster. They end up taking a large dose of insulin in the morning, then coast along on that one shot of the hormone for the whole day. The dose is based on the individual patient's normal rate of sugar metabolism and on estimates of what his food intake will be for the day. The unnatural thing is that such a regimen provides an abnormally high insulin concentration in the morning, a little less at noon, and a low concentration at night. Ideally, however, insulin should be supplied in precise, steady doses exactly when needed by the body, duplicating as closely as possible the way normal beta cells would respond.

What this suggests, then—if you're looking to better manage or even cure diabetes—is that several different approaches might be possible, including even genetic engineering—someday.

Perhaps the most likely and first approach—which will actually be a better form of diabetes management instead of a real cure—will be to build a tiny electromechanical system that will monitor the body's need for insulin, then turn on a small pump to deliver a precise amount of the hormone into the bloodstream. Attached to the egg-sized pump would be a year's supply of insulin inside a small reservoir. Dr. Samuel Bessman, at the University of Southern California School of Medicine, has already developed and built such a blood-sugar sensing system, pump, and reservoir to keep control of the blood's sugar supply. Bessman's little sensor and pump system is almost ready to be implanted into patients, once approval is received from numerous sources, including the U.S. Food and Drug Administration.

A second approach to diabetes and one which might also be used to deal with other genetic ailments—might best be called a step into the world of cellular engineering. This would involve taking a group of beta cells from a person whose tissues closely match the patient's, then implanting them onto the diabetic's pancreas. Presumably, if the beta cells are in proper condition, they will then respond to the patient's interior chemical signals and pump out plenty of insulin right on cue. This is, of course, a large assumption, since the problem of diabetes may lie instead with the body's inability to give the right signals to *any* beta cells, healthy or not. Thus, if the body's cells don't respond to insulin properly, even vast numbers of new beta cells might not do much good.

Furthermore, even if the newly implanted beta cells are closely matched with the patient's own body chemistry, chances are still strong that this new foreign tissue—these new "unself" foreign cells—will be rejected by the patient's protective immune system. Thus the beta cell transplant method may someday offer a good choice for combating diabetes, but significant problems remain unsolved.

The final approach to curing diabetes would represent true genetic engineering. This would involve taking some of the patient's own beta cells, growing them up in a petri dish, then altering them through genetic tinkering to become normal insulin-producing cells. If one or more of the genes inside these cells are defective and make the cells unable to produce insulin in proper amounts, that gene—or genes—could be changed or exchanged for a good copy. Then the new, repaired insulin-producing cells would be transplanted—actually reimplanted—into the diabetic.

The approach of using the patient's own cells represents a way for maneuvering around the body's natural tendency to reject foreign tissues. The answer would thus lie in giving the body some of its own tissues rather than tissues from someone else. It is thus assumed, perhaps too naïvely, that such genetic tinkering would do nothing to change the original sample of cells into new types of cells that the body would later not recognize as "self."

It has recently been suggested, too, that diabetes may join the growing list of ailments known as autoimmune diseases. These maladies involve the body's immune system going astray, attacking some of the patient's own cells, even though these cells were formerly seen as "self." In diabetes, indeed, there is some new evidence that the patient's own powerful killer lymphocytes may be the cause of the disease, suddenly turning against the beta cells, putting them out of business, shutting down the normal supply of insulin. Scientists aren't sure yet what might cause this autoimmune response, but a few have suggested that a virus may come along, attack the beta cells, suddenly making them appear "foreign" to the lymphocytes.

On the other hand, if it turns out that the real cause of diabetes is the body's imperfect response to insulin, even genetic engineers will be hard pressed to find ways to make a huge number of a patient's cells respond better to insulin. This task may be too difficult, after all, to even warrant trying, but one can still suggest that some specially designed viruses—carrying just the right genetic information—might be used as messengers to carry new genes into many of the living cells.

Since that's exactly how a virus operates—by invading cells and inserting its own genetic information, presumably into the chromosomes—a virus would seem to be the ideal, but possibly dangerous, messenger for transmitting genetic code sequences into living cells. This shouldn't suggest, however, that a doctor can overlook the dangers involved, that he can go around infecting people willy-nilly with viruses just for the sake of experiment.

Nonetheless, this approach—using a virus to give a disease victim a few new genes—has already been tried on three little girls in Germany, but the attempts were unsuccessful. According to a detailed report in *Medical World News* magazine, in 1970 a German physician—Dr. Heinz George Terheggen, at the Municipal Children's Hospital in Cologne—tried to use a strange virus called the Shope papilloma virus, which causes huge horny warts to grow on the skin of Kansas cottontail rabbits, to combat the three girls' inborn metabolic disease, hyperargininemia. This ailment is characterized by the patient's inability to produce one important enzyme, arginase, allowing the amino acid arginine to accumulate in the blood and spinal fluid. Accumulation of arginine interferes with brain function and muscle control, causing seizures, spastic palsy, liver problems and vomiting. And, as in most other genetic diseases involving enzyme deficiencies, this problem is inborn, irreversible, and grows progressively worse.

After discovery of this new genetic disease—in the first two of those German girls—was announced in 1968, it was suggested to American research specialist Dr. Stanfield Rogers at the Oak Ridge National Laboratories in Tennessee that Shope papilloma virus (SPV) might be of some use against hyperargininemia. Early experiments done by the discoverer of the virus, Dr. Richard E. Shope, indicated, indeed, that SPV carries the genetic information for construction of arginase, the missing enzyme, and that when SPV is injected into humans—including Shope himself in 1933—the only noticeable effect is a decline in the blood's arginine level. Too, a similar effect was found among laboratory workers who had merely been exposed to the virus.

These results, then, prompted Rogers, in Tennessee, to offer live samples of the virus to Terheggen in Germany. The hope was that if Terheggen chose to do the experiments, SPV might similarly reduce the arginine levels in the stricken girls. What this amounted to, of course, was the first attempt at direct gene therapy, the first attempt to insert a gene in a living patient's cells with the hope the new bit of genetic information might be made to work. Thus, in May 1970, Terheggen went ahead and injected the virus into two girls. Unfortunately, the levels of arginine in their blood remained unchanged.

Only a year later, however, and despite the advice of doctors, the parents of these girls went ahead with another pregnancy, again bearing a baby girl with inborn hyperargininemia. The parents then declined to allow the child to be hospitalized, so it was only much later that she was eventually inoculated with SPV. As in the other cases, this, too, failed.

Why didn't it work?

Both Terheggen and Rogers believed that the virus injected into all three girls must have deteriorated during the long storage period in Germany, and they probably confirmed this by trying SPV against its normal host, the rabbit. Again, no response.

These researchers did say in 1975, however, that plans were being made to give a second dose of SPV to each girl, even though there is no chance that any improvement in their mental retardation is possible. The goal of the new experiments will be to find out whether this type of therapy might work at all. If it does, then SPV might someday be used in other cases of hyperargininemia, especially if the disease can be caught and treated right at birth, before the damage has occurred.

While this might seem to be one hopeful prospect against a terrible genetic disease, there also appears to be ample reason to argue against these experiments, at least for the present. Other doctors and research scientists warned, strongly, that Terheggen and Rogers were playing with dangerous stuff, since the Shope papilloma virus is also known to cause tumors in rabbits, and might conceivably lead to cancer in human beings. The argument turns on the point that it's not worth further risk to these three girls, especially when no improvement in their condition seems possible. Thus it boils down to one difficult choice: leave the patients alone, letting them go through life hopelessly incapacitated, or inject a virus that might help cure the genetic disease, but which might also lead to cancer twenty, thirty, or forty years down the road. It's too late, of course, to really help those three German girls, but what about children in the future?

More recently, however, in less ambitious, better-controlled experiments, Dr. Carl R. Merrill at the U.S. National Institute of Mental Health (NIMH) has been able to correct a genetic defect through use of the viral messenger method—using a live virus as the gene carrier—but only in a sample of some living human cells growing in a laboratory dish. These cells had been taken from a

patient born with galactosemia, and Merrill used a special virus—
the Lambda virus, a favorite tool for biologists who want to tam-
per with the genes—to infect the cells. The genetic information
needed to correct a deficiency in galactose metabolism had been
inserted into the virus's DNA by having it first infect the bacterium *E. coli* to pick up the missing genetic instructions.

Merrill and his coworkers—Mark R. Geier and Dr. John C.
Petricciani—then induced the virus to infect cultured skin cells,
and went on to prove that the cell's protein-building machinery
was able to read the virus's genetic messages and act on them to
produce the missing enzyme. What this NIMH group didn't prove,
however, was whether these new genes—which code for the crucial enzyme—are actually intermingled with the cell's original
supply of DNA, or whether this information is somehow being
used, being read and transcribed, without becoming part of the
cell's own genetic codebook. It was shown, though, that the activity of these special genes does continue even after the infected
cells have divided several times, indicating that the reproduction
of these genes is somehow occurring.

An interesting point too, Merrill noted, is that this represents
the first time the Lambda virus has been induced to invade human
cells, and it is also the first time that genes that came from a bacterium have been made to work so well inside the cells of a
higher animal. The Lambda virus normally infects only bacteria,
and for this reason is known as a bacteriophage. When biologists
and biochemists speak of doing "phage work," they're discussing
the use of such viruses to study the genetic properties of both
bacteria and viruses.

The work by Merrill and his colleagues at NIMH certainly suggests that this promising technique may soon begin finding widespread use, in experiments, if not yet in patients. It seems possible,
indeed, that some bits of cultured human tissue—once they begin
using new genes to produce the proper products—might someday
be considered for reimplantation. If such a technique works, it
could represent the realization of a longstanding goal—the actual
cure of a genetic disease.

Unfortunately, use of viruses for infection of human cells may present hazards that doctors and their patients aren't willing to face. It is well documented, for instance, that some viruses carry genes capable of transforming normal cells into cancer cells. Thus, while a virus might be used to carry important genetic information into a cell, it may inadvertently carry along a few unwanted messages as well. This would suggest that much more research may be needed before this technique becomes a real option for the treatment of disease in humans.

However distant the use of viruses or other genetic engineering techniques may seem, let's remember that medical science has already gone a substantial distance toward managing a few of the many genetic diseases, at least in crude ways.

Indeed, biologists and other specialists have learned enough in the 130 years since Gregor Mendel discovered the science of genetics to know, at least, what *not* to do. It is certainly a well-documented principle, for example, that matings between close relatives such as brother and sister should be forbidden, since such pairings often lead to genetic disaster for the offspring. Doctors have also learned to discourage mating by people who obviously show they're carrying similar genetic diseases or defects, such as two people with sickle-cell anemia, or those with the genes for Tay-Sachs disease. This doesn't mean these people can be kept from pairing, just that we do realize what the consequences may be.

Because this knowledge of genetics has been developed, and because of some demand for its use, many medical centers have been setting up genetic counseling programs to help people make these "go" or "no go" decisions. Couples who suspect one or both may be carrying a hidden genetic defect which might be passed on to their child if they mate can seek advice and actually check out the odds. Every pregnancy, of course, represents a gamble, but genetic counseling and testing have provided a new way to check the dice before they're thrown. It's possible now, in many cases, to tell a couple what the precise probability is—for or against —of having a genetically defective child. This means they'll have much more information on which to make important decisions and, as time moves on, this quoting of the odds on genetic problems will find more and more use as the techniques improve.

As mentioned earlier, steps have also been taken that carry doctors and their patients much beyond genetic counseling. In the past few years, methods have been found which allow physicians to even test the genetic health of the child—for some diseases, anyway—before it even leaves the womb. If these tests show that an unborn child is obviously defective, then the pregnancy can be terminated; an abortion can be performed.

Of course some people are still balking at the whole idea of abortion, but it is nonetheless one effective method for eliminating defective gene combinations when they occur in the population's constantly shuffling genetic deck of cards. The difficult question of morality—in relation to abortion—will be argued for years yet to come, and there will probably never be a wholly satisfactory answer.

Nonetheless, the techniques are now well established for performing abortions, and when this is combined with the newer methods for assessing the genetic health of the unborn infant, abortion does provide an important, powerful new tool for combating genetic ailments.

The procedure of testing the living fetus for genetic abnormalities is called amniocentesis. Basically, this technique involves a long, hollow needle—inserted directly through the abdominal wall, through the uterine wall and into the amniotic sac—to sample the watery amniotic fluid that surrounds the child. This is still a difficult, touchy procedure, and doctors warn that it should only be done when there is absolute need, not for mere curiosity about trivial things like finding out the sex of the unborn child.

The small sample of amniotic fluid can be used for several types of tests. Scientists and doctors usually examine the few living cells in the amniotic fluid that have been sloughed off the child. What doctors look for depends on which diseases are suspected. If, for example, the child is being carried by an older woman, a woman who was near age forty at the time of conception, her physician would be looking closely at the chromosome arrangement in the cells found afloat in amniotic fluid. It is well known that mongolism—or Down's syndrome—is a defect involving the presence of an extra twenty-first chromosome. And mongolism is strongly associated with pregnancies in older women. Thus, if a doctor suspects a mongoloid child, the amniotic fluid can be sampled and the cells, after being grown in culture, can be tested for the chromosomal marker associated with mongolism. The clear sign —the extra chromosome—tells a physician that the child is definitely going to be mongoloid.

Once the tests have been made and the results have been read, however, it will then be up to the parents—on their doctor's advice —to decide whether to terminate the pregnancy or to go ahead with full knowledge they'll be caring for an abnormal child; a child who may live to be an adult who is unable to cope with the demands of society.

In cases of mongolism, this burden is often balanced, on the other hand, by the prospect of having a very responsive, loving child who thrives on close relationships. Some parents of these children have found, indeed, that their lives are filled with love and caring because of the simple, trusting nature of a mongoloid child. Some doctors report that for some parents it has been a very rewarding experience in spite of the very obvious problems.

It is conceivable, however, that sometime in the future the abortion decision won't be left to prospective parents and their doctor alone. Some scientists, looking perhaps very far ahead, suggest that such burdened children, the mongoloids and those with other genetic problems, will someday not be allowed to be born at all, under any circumstances. It may be decided—perhaps in some "brave new world"—that defective babies are too expensive, both in terms of the time and the money involved in caring for them. Thus society will be deciding, perhaps unjustly, it cannot afford sixty years of caring for and feeding individuals who will never, ever, become productive citizens. Such suggestions may sound cruel, perhaps even inhuman and immoral, but even a cursory look inside any of today's state-run hospitals for the mentally retarded will show why the idea is considered attractive and even humanitarian by some.

As biologist James Bonner, at the California Institute of Technology put it: "Someday, when we finally realize that nobody can have more than two children, it will then make sense to have only the very best two children you can possibly have," which suggests that society won't allow babies with obvious genetic defects to be born.

Detection of mongolism isn't the only use for this new tool called amniocentesis. Biologists and doctors have found that it is difficult, but not impossible, to take a few of those live cells from the amniotic fluid, isolate them, and culture tissue from them. Thus in addition to simply looking for obvious chromosomal abnormalities, it is also possible to test for defective genes by checking to see if they're failing to produce specific enzymes. Eventually, as techniques improve, it shouldn't be difficult to test the whole genetic makeup of the unborn child instead of just looking for a very few specific metabolic abnormalities, as is being done now. What it means is that a new window has been opened up that lets the doctor inspect the child, in biochemical terms at least, well before birth. Now, if a family is suspected—or itself suspects—that a particularly bad set of genes has been dealt, such a combination can be spotted and eliminated well before the child is born. This can be done with enough time left for legal abortion.

Ideally, however, the long-term goal of the medical profession will be to cure the child of genetic defects rather than just terminate the pregnancy. As a result, what genetic engineering will finally offer will probably be new ways to replace defective genes, giving every child—if properly diagnosed in advance—a better chance to live a full, fruitful life.

Where are we now?

In America, where the diseases caused by infection and by environmental hazards have been increasingly brought under control, the remaining genetically based ailments have been standing out with greater and greater prominence. With polio virtually eliminated, for example, the March of Dimes has changed, becoming the National Foundation-March of Dimes, and is concerned now mostly with birth defects such as genetic diseases. So now diseases caused by chromosomal problems, by unfortunate combinations of genes or by undesirable mutations, have become more important and more visible as they have come from behind the screen of other devastating, more prevalent diseases.

National figures tell us that little real progress has been made toward the control and treatment of genetic diseases, except in cases where the problem can be manipulated by diet or hormone injections. The death rate from genetic diseases—mainly caused by congenital defects or metabolic imbalances—has fallen only slightly in the twentieth century. In 1900 only one in thirty infant deaths was caused by genetic problems. Now, however, the number is one in five. This, of course, reflects the general decline in infant mortality, which has been spectacular. But it also serves now to magnify the presence of genetic ailments. As a consequence of good medicine, they have become just that much more visible.

Specialists who have achieved many of the advances mentioned in the care and treatment of genetically damaged individuals have also found, through laboratory studies, that inherited abnormalities can be divided another way, into three groups, on what might be considered—by laymen at least—rather superficial grounds. First are the abnormalities caused by abnormal numbers or shapes of chromosomes, which lead to maladies like mongolism. Second are those caused by mutations of individual genes which are passed on to offspring, such as metabolic disorders like PKU and galactosemia. Third are the ailments such as heart disease, cancer, emphysema, and others which are probably attributable to a complex and still mysterious mix of genetic and environmental factors.

In that first classification, the chromosomally based disorders, most of the errors are caused by mistakes occurring in the process of meiosis—that special cell-division mechanism that produces the individual sperm and egg cells, each carrying half of the individual's chromosome supply. As a group, these abnormalities can then be further subdivided into diseases involving the sex chromosomes—those X and Y chromosomes—and all the other chromosomes, which are collectively called the autosomes, or nonsex chromosomes. As an illustration of how important chromosomal errors are in the whole picture of genetic diseases, recent studies have shown that more than half of all spontaneously aborted fetuses are truly abnormal, unable to live because of scrambling of these tiny chromosomes. The most frequent genetic abnormality (51 percent) seen in humans is the presence of one extra chromosome. Thus Down's syndrome, mongolism—which is caused by the presence of an extra chromosome—represents the most prevalent genetic abnormality in man. The frequency now is about one child out of a thousand.

It should be recalled, of course, that the normal pattern in humans is for every cell to contain twenty-two pairs of autosomal chromosomes plus the pair of sex chromosomes, either XX (female) or XY (male).

As noted earlier, mongolism results from the new individual's being dealt one extra chromosome, an extra or third copy of chromosome number twenty-one. What hasn't been explained yet is why the risk of having a mongoloid child increases substantially as the age of the mother increases. Indeed, this increase in risk eventually reaches 1 percent in women who become pregnant after reaching forty. Less well known, but also caused by chromosomal aberrations, are Edwards's syndrome, which involves the presence of an extra eighteenth chromosome, and Patou's syndrome, in which the patient is born with an extra thirteenth chromosome. Another similarly caused autosomal abnormality is called the "cat's cry" syndrome because of the distinctly abnormal sounds the baby makes. This is caused by part of chromosome five being missing.

The other category of chromosomal abnormalities—those caused by defects in the sex chromosomes—are most often caused by the presence of one or more extra sex chromosomes. There are also some rarer abnormalities caused by a sex chromosome being missing. In one of these cases, the patient is born carrying only one X chromosome—a condition known as Turner's syndrome—and fails to develop secondary sex characteristics at the age of maturity. In addition, these patients are also burdened with distinctive physical abnormalities such as webbing of the neck.

Also well known by specialists are cases in which a person is dealt multiple sex chromosomes, such as three X chromosomes. This is a situation known, logically, as the triple-X syndrome. Other abnormalities are the XXY condition known as Klinefelter's syndrome, and the XYY combination.

Extra sex chromosomes recall the notorious case of Richard Speck. In 1966 Speck was accused and later convicted of breaking into the Chicago apartment occupied by a group of nurses, killing eight of them by stabbing and strangulation. Later, as his case moved toward appeal, it was reported that Speck is one of those rare individuals born with the XYY syndrome, a condition suspected of making men overly aggressive. Studies indicated that the overly aggressive inmates in penal institutions are about sixty times more likely to carry the extra Y chromosome than men in the general population. At that time, some 100 men with this genetic abnormality had been discovered, and these included at least four who had been charged with or convicted of murder.

That spectacular murder of eight nurses, the tall, acne-scarred defendant, and the strange reports of a mysterious genetic abnormality generated great attention for the whole subject of genetic problems, but as it later turned out—after more extensive laboratory tests—Speck was not a carrier of this chromosomal abnormality. He had been wrongly diagnosed, and the furor over the XYY "supermale" condition was, for Speck, at least, baseless.

Genetic markers can be very good tools, if they are used right and read properly. But the Speck case also points out that there's much yet to be learned about genetic problems and their effects, especially their effects on behavior. There is still much argument going on about what behavioral problems, if any, arise from having an extra chromosome, what form such problems might take, at what ages they might show up, and how they might be passed on, if at all, from one generation to the next.

Despite the errors of the Speck case, however, the important lesson coming from more recent work in genetic diagnosis—specifically, from amniocentesis—is still that such abnormalities can be seen, ever more accurately, well before a damaged child is born. It also still means that families facing this problem must choose what they'll do about such a diagnosis. Thus it still presents that same moral dilemma: whether to seek an abortion. It's obvious, of course, that there are no cures yet for these abnormalities, so once the child is born—if born—the question remains about what role he might be able to play in society.

Unfortunately, it will probably be up to society eventually to decide if individuals with abnormal sex chromosomes are to be singled out and watched, kept under surveillance as potentially dangerous because of overaggressive tendencies. Chances are strong that society, in many countries, will decide in favor of surveillance; but it is predictable that serious opposition will be voiced to the idea of tagging persons as "inferior," or "troublemakers" merely on the basis of their genetic makeup. Critics are already pointing out that such people will be effectively branded for life by genetic diagnosis, regardless of whether they will ever commit a violent crime, or any sort of crime. There is new evidence, too, that an extra Y chromosome may not, after all, predispose a man to be overly aggressive or violent.

Whatever the outcome of this bitter debate, one great temptation—given modern genetic diagnostic tools—is to set up wide screening programs to begin searching for persons carrying abnormal genetic combinations—caused either by strange chromosome count, by mutations, or by unfortunate combinations of the genes. These screening programs, perhaps best illustrated by recent efforts to search out carriers of Tay-Sachs disease and sickle-cell anemia, have tended to create their own unique sets of problems, and some screening programs have proved not to be very successful. Nonetheless, a lengthy report published in *Science*, the magazine of the American Association for the Advancement of Science, reported that a majority of American states has already enacted laws requiring some sort of genetic screening, usually involving blood tests on infants immediately after birth, looking primarily for PKU (phenylketonuria), a metabolic disorder which causes mental retardation unless treatment begins within the first few weeks after birth. Currently some 90 percent of the newborn infants in the United States are being screened for PKU before they leave the hospital. In some cases, parents can block these PKU screening tests, but most parents don't seem to be aware that such tests are being made in the first place.

Of course, PKU testing is merely the entering end of a larger wedge, since it requires only small legal steps to arrange for ever more comprehensive testing programs that would screen for a whole range of genetic abnormalities, some of which might not really be considered genetic ailments at all. Too, with continuing improvements in computer-controlled, automated analysis of blood chemistry, the temptation to require more screening is even greater as speed increases and costs come down. A further step that can be foreseen would be to make complete genetic screening mandatory at birth for every baby, perhaps thereby opening up every child's genes—its entire genetic repertoire—to public scrutiny. Whether this is going to be desirable will be open to serious question, but whether it is going to be possible is hardly even debatable. It can almost be done now.

Despite all the arguments over mass screening and whether it's necessary, the trend does seem to be clearly heading in the direction of more and more screening. According to *Science,* many of the states are requiring testing for additional diseases, adding to the list simply by adding new maladies to be looked for along with PKU. Critics point out that the U.S. Congress made a mistake doing this, however, when it began creating special programs for genetic diseases, first a sickle-cell anemia program, next a program for hemophilia, and on and on—amounting to what *Science* called a "disease-of-the-month approach."

Often the diseases selected by some states as candidates for screening programs are relatively rare, and in some cases shouldn't be considered suitable for screening anyway. Some of these genetic diseases are still very poorly understood. In others, whatever information might be gained by screening is of doubtful value, and in still other ailments there's nothing that can be done anyway—other than perhaps seek abortion.

Some of the people being screened may end up being hurt in one way or another. In some cases those identified as having genetic ailments—such as blacks with sickle-cell anemia—have felt stigmatized when told there's something wrong deep inside their genetic deck of cards. The question all this raises, of course, is whether some mass screening programs are really worth the effort. The answer is that nobody really knows yet.

Perhaps the best approach yet to mass genetic screening will come from passage of a new federal law, a national Genetic Disease Act creating a special unit within the U.S. Department of Health, Education and Welfare (HEW). Such a group would provide support for research, for better personnel training, and for public education programs concerning genetic disease. This shouldn't say, however, that the U.S. government will necessarily be taking over the whole genetic screening business. More probably, individual states will be increasing their own efforts to make genetic screening more available, and in some instances even mandatory.

Obviously, then, the spread of PKU testing to more than forty states serves as a good example of how screening programs can be set up, evolve, and spread.

PKU testing itself, however, isn't really a brand-new procedure. Scientists and doctors were able to develop the first simple, inexpensive test for PKU as early as 1961. The test merely involves taking a few drops of blood from the baby's heel, then analyzing it to see if an excessive amount of one amino acid, phenylalanine, is present. This amino acid is important in a normal child's developmental processes, but in cases of PKU the baby's metabolic system lacks an important enzyme needed for chopping up the phenylalanine molecule. As a result, phenylalanine begins accumulating, and it eventually causes damage to the infant's central nervous system, including the brain. This, of course, would help explain why PKU children suffer severe mental retardation.

Treatment of this condition—which must begin early to avoid irreversible brain damage—involves simply putting the children on a low-phenylalanine diet, essentially a protein-free diet, until they are from four to six years old. This relatively simple step has been successful in saving a number of children from mental retardation, keeping many of them from spending their lives locked up in mental institutions. Obviously, state legislators, when faced with an inexpensive way to save some children from retardation

—and at the same time save the millions of dollars which would be spent caring for them—have no real choice but to vote approval for PKU screening. Indeed, the costs of PKU screening are minuscule when compared to what it costs to keep even one child fed, clothed, and housed in a mental hospital for life.

Of course, work with genetic diseases didn't just stop with the one success of PKU testing. While legislators were busy setting up screening for PKU, medical research teams moved on to work with other genetic maladies, developing additional simple tests capable of detecting abnormal concentrations of important amino acids either in the blood, in spinal fluid, in amniotic fluid, or even in the mother's urine. Almost simultaneously, the ability to identify abnormal chromosomal combinations was being worked out, and doctors suddenly found they were able to screen adults for things like sickle-cell anemia and Tay-Sachs disease, babies for abnormalities like PKU, and unborn children—fetuses—for inborn errors of metabolism like galactosemia or chromosomal abnormalities that cause mongolism. Little wonder, then, that the temptation to test for almost everything became hard to resist.

A good example comes from New York State, where all babies born in the past few years have been screened for PKU, for sickle-cell anemia, for maple-syrup urine disease, for homocystinuria, adenosine deaminase deficiency, and histidinemia. The annual cost to the state has been something like only $250,000. When this amount is compared to the cost—the whole lifetime cost—of caring for only a few retarded children, testing programs seem to be an honest bargain. This is especially true when there is a good chance that some of these children, such as those born with PKU, can be treated successfully.

There are nonetheless some valid arguments on the other side of the genetic screening debate, even though they appear hard to balance against the success being found in some screening programs. The arguments still deserve airing, however, and are perhaps best illustrated by the screening programs begun for identifying persons carrying the genes for sickle-cell anemia. The critics point out, first, that there is really no point in searching for victims of sickle-cell anemia, even among babies, since there is nothing yet that can be done to help them, and it's a disease that doesn't show up until later in life anyway.

The second point critics make is that by identifying babies born with sickle-cell trait, society also identifies their parents as sickle-cell carriers. The carriers, however, normally show no obvious symptoms of the disease, and as a genetic ailment it can cause problems only when two carriers mate and produce a child who gets dealt two genes coding for sickle-cell anemia. Since a few people believe that to be branded as a carrier of sickle-cell trait is tantamount to being publicly named inferior, the screening programs could end up doing as much harm as good, in a few cases.

An interesting sidelight about sickle-cell anemia is that in tropical Africa, where the trait apparently originated, this sickling of the red blood cells characteristic of the disease appears to confer a definite advantage on its victims, since it tends to protect against malaria.

Another point about the uses of genetic screening, and about the Western world's growing preoccupation with genetic ailments, comes from Dr. Sydney Brenner, of England's Medical Research Council laboratory in Cambridge. Brenner pointed out in an interview that medicine should be viewing its problems, its challenges on a world scale, from the standpoint of serving everyone in the world. If the problems of medicine are seen from this point of view, the different diseases take on a completely new perspective, and the genetically based ailments become obscured. Brenner insisted, indeed, that in this larger context the diseases that worry Americans—such as childhood leukemia—seem almost trivial; they hardly exist.

"What I mean," Brenner noted, "is that there are a million people still at risk from malaria. It's true, malaria has been wiped out in the West, but the question is, still, where do we apply our medical talents?"

Despite this continuing worldwide medical imbalance, a large proportion of the research done in the United States and in other advanced nations is still focused on the prominent home-grown diseases, on heart disease, cancer, and, of course, genetic diseases. Indeed, now that amniocentesis has increasing reliability, some experimenters and a few doctors are hoping to find ways to treat genetic diseases well before a child is even born. Already, in some special cases, as soon as a child is discovered to have a genetic disorder—even while still in the womb—there are some treatments that can be done successfully. One case, as reported in the *New England Journal of Medicine*, involved diagnosis of an inborn metabolic disease called methylmalonic acidemia. The treatment —which turned out to be successful—merely involved giving the mother large doses of vitamin B_{12}.

Despite this instance of success, however, chances that very many babies can be treated for genetic abnormalities in utero (in the womb), before birth, appear to be minimal. According to the pessimistic view of Carlo Valenti, at the Downstate Medical Center in New York City, "Chances that useful treatment [of unborn children] will soon be at hand are slim. For many conditions, irreversible anatomical changes are probably present so early in gestation that their prevention or reversal in utero by drug or enzyme treatment seems highly improbable. . . . Although I would welcome an alternative to abortion of a defective fetus, I reluctantly conclude that abortion must remain the solution to inheritable diseases."

If this is the case, if abortion is to remain the most useful response to genetic diseases in the foreseeable future, then the process of amniocentesis is going to become a more vital, more common test.

Indeed, geneticist Tabitha Powledge writes in *New Scientist* that amniocentesis has moved rapidly from being a mere experimental procedure to become a clinical diagnostic technique. This, she said, occurred semiofficially on October 20, 1975, during a special symposium in Washington, D.C., sponsored by the U.S. Department of Health, Education and Welfare. The occasion was formal release of a study done by the National Registry for Amniocentesis, compiled with the aid of nine large medical centers and coordinated by the National Institute of Child Health and Human Development.

Involved in the study were 1,040 pregnant women who had undergone amniocentesis. Results indicated that only 3.5 percent of these women suffered spontaneous abortion or delivered still-born children. This was compared to a rate of 3.2 percent in a matched control group of 992 women who had not undergone amniocentesis. Also, careful examination of the babies at birth, and again at age one, showed no signs of injury—either direct or indirect—from amniocentesis.

Perhaps the most important finding from this nationwide study of amniocentesis was that the reliability of fetal diagnosis is really quite high, 99.3 percent accurate. Out of a total of only six errors, two involved misdiagnoses of children who were later born mongoloid; three errors were made in attempts to determine the fetus's sex; and one child was misdiagnosed as having galactosemia. Fortunately, this infant's parents decided against abortion, since the disease—had it existed in this child—could have been treated successfully after birth through manipulation of its diet.

Nonetheless, even though the reliability rate for amniocentesis was found to be high, Ms. Powledge commented in *New Scientist* that reliability may still be a problem in some cases, especially when tests involve laboratory examination of the baby's cells. One problem is that it takes a few weeks to get these cells to grow in culture, and they sometimes die before tests can be made. In the federally sponsored study of amniocentesis, for example, 13 percent of the women were required to undergo the fluid-sampling procedure twice, most often because the original sample of fetal cells refused to grow.

Taking that second sample, too, is considered more hazardous: first, because it reimposes all the risks associated with the first attempt; second, because the additional time needed for growing up the tissue sample, for testing it, and then for making the decision may mean—if an abortion is sought—that the operation will be considerably more dangerous. Indeed, it could also mean that the legal deadline for having an abortion has passed.

What is the real need for amniocentesis? Does this rather exotic procedure deserve all the emphasis it's getting? Some estimates, indeed, show that only 300,000 women in the United States are, at any one time, potential candidates for amniocentesis because of their age, family histories of genetic ailments, or their prior experience of having already borne an abnormal child. Even for this small portion of the pregnant population, too, the facilities and manpower for amniocentesis are not available. Furthermore, even if they were available, there is little certainty they would be used. This was exemplified by a Massachusetts study done in 1974, showing that less than 5 percent of the pregnant women over age thirty-five took advantage of amniocentesis even though the procedure was available and they represent the group most "at risk" for having mongoloid children.

Also, the cost of amniocentesis is not trivial. At present the procedure costs from $150 to $200, and the price is still going up. In many past instances, the cost of amniocentesis was subsidized by federal grants for research, so it was done partly as an experimental procedure. Thus for the near future, women seeking amniocentesis will encounter a sizable expense atop the already-high cost of having a baby. It should thus be obvious that as amniocentesis becomes more widely known, more available and perhaps somewhat safer, it will probably be the affluent who take most advantage of such tests—unless it is somehow subsidized by government, as is being done in some states. Nebraska and Tennessee are already offering free amniocentesis, along with the associated cell culture studies, to all pregnant women over age thirty-five. Similar access is also being considered in New York, where the state may yet wind up having an extensive prenatal diagnosis program.

Discussions leading in that direction, held at Columbia University, included proposals that would have all pregnant women over age forty, at least in the beginning of such a program, be offered amniocentesis. If adopted, this program would progress in four distinct steps until all women in the state would be screened during pregnancy using amniocentesis. Indeed, it was calculated that by testing only women over age forty, some 13 percent of the cases of mongolism could be prevented through abortion. Later, in this program's final stages, with almost all pregnancies subject to prenatal screening, as many as 90 percent of the potential mongoloid children could be identified before birth and could possibly be aborted.

However, some observers from the medical profession warn that such a diagnostic system might be wide open to abuse. As a way of determining the sex of a child before birth, amniocentesis is already providing strong temptation for couples to seek abortion for reasons other than health of the child. Worse, not only might this lead to frivolous use of a dangerous procedure, it's already been reported that some prospective parents, disappointed at learning they're not having exactly the child of their choice, have chosen to end the pregnancy on the basis of sex alone. Perhaps, then, it should be asked whether sex-finding is a responsible use of amniocentesis, and whether the presence of a child "of the wrong sex" is an acceptable reason for abortion. In areas where abortion is available on demand this seems to be happening already. Unfortunately it will probably soon all boil down to a continuing, shrill, and unreasonable debate. Clearly, all these questions about amniocentesis won't be fully answered until the argument over abortion is resolved, if it ever is.

Meanwhile, scientists and doctors are moving ahead, working on new developments that may make some of these arguments obsolete. One newer technique, which may be used instead of amniocentesis, involves the use of ultra-high-frequency sound equipment to make a sound-picture of the child in the womb. In some applications, ultrasound is being used as a way to locate the placenta in the womb so that amniocentesis can be done more accurately. Enthusiastic researchers are also predicting that it will soon become possible to make three-dimensional studies of the unborn child, and it may also become possible even to study individual organs inside the fetus.

A good example of what's possible through this ultrasound approach comes from the work of Dr. Jason Birnholz of Stanford University, who reports he can—80 to 90 percent of the time—successfully predict the sex of a fetus as far as two months ahead of birth. Birnholz said he has been testing an advanced ultrasound scanner made by G. D. Searle, and once an operator knows what to look for, determining an unborn child's sex is quite simple.

One of the important things about ultrasound—and one point where it differs from amniocentesis—is that ultrasound requires no physical violation of the amniotic sac. Imaging is done by carefully reading the sound waves reflected off bones, tissues, and fluids inside the uterus.

Basically, ultrasonics uses high-frequency sound waves that are beamed through the body. The echoes produced by objects hidden inside are picked up by electronic listening devices—supersensitive microphones. They are fed into a computer, unscrambled, and then displayed on a television screen as a pattern of tiny dots. Thousands of these echo dots combine to yield a reasonably accurate image. The sound waves themselves are not harmful to the mother or the baby.

In other medical applications, ultrasound equipment is being used to study living organs in adults, especially for pinpointing where damage has occurred in the heart muscle after a heart attack.

Another instrument that is becoming useful in diagnosing genetic and birth defect problems afflicting the unborn child is the fetoscope, a device which actually allows the doctor to look inside the womb. The fetoscope uses thin optical fibers—which themselves are still being developed and improved—to pipe light into the uterus and then carry the image of the fetus out. Abnormally formed limbs can thus be seen directly, and malformations like spina bifida—which is a congenital defect in which the base of the spine fails to close during development, leaving an open wound through which spinal fluid can leak—can also be seen. In spina bifida, the infant is often paralyzed and mentally retarded. Amniocentesis can detect this malformation because spinal fluid is present in the amniotic fluid.

So far, of course, emphasis here has been on diagnosis of genetic illness in the fetus or the newborn infant. Work is also under way aimed at adults and older children, toward diagnosis, treatment and—hopefully—someday the cure of inborn abnormalities. Diabetes—whether inborn or not—was discussed before as one good example of such work, but there are others as well.

Another example is the research being done that might help correct a severe genetic deficiency which causes some children—in rare instances—to be born without immune systems. Without this important defense mechanism, the individual's body quickly succumbs to disease organisms. These children seldom survive. Now, however, Dr. Robert Good and other research specialists are working on ways to activate the immune response through manipulation of cells, such as implantation of bone marrow, and in work on glandular tissue. Such work has indeed apparently led to immune capabilities being established in some children.

Rather than the implantation of new bone marrow tissue, another approach tries to plug in some new healthy genes. Dr. Michael D. Garrick at the State University of New York in Buffalo has been experimenting with the possibility of transplanting normal genes for the production of red blood cells—specifically the genes for hemoglobin—into the bone marrow of patients with hereditary deficiencies. Working under a grant from the National Foundation/March of Dimes, Garrick is aiming his efforts specifically at sickle-cell disease and thalassemia.

Garrick explained that attention is now being focused on how efficiently the new normal genes are incorporated into the patient's tissues, what harmful side effects might arise, and whether or not the new genes are transmissible from one generation to the next.

Similar studies with bone marrow are also being used as one approach to finding a cure for leukemia. Research under way at the University of California, Los Angeles, has been limited so far to small animals—primarily mice—but the work does suggest that bone marrow transplants might be effective against this form of cancer. Leukemia occurs when cells of the bone marrow, the lymph nodes or other blood-forming tissue—which serve as factories for renewing the blood's supply of specialized blood cells— run wild to produce abnormally large amounts of specific types of blood cells, usually white blood cells.

Currently the best, most effective treatment for leukemia appears to be the use of anticancer drugs, which are administered over precise periods of time in hopes of killing all of the rapidly multiplying cancer cells, while allowing normal cells to survive. The basic problem lies in assuring that every last renegade cancer cell is destroyed, since each malignant cell can act as the seed for restarting the disease over again. Now, after years of trying to combat leukemia with ever more potent, more versatile drugs, it has finally been found that an array of anticancer drugs, working in combination, can be given together in carefully timed doses so that even fewer, if any, of the cancer cells survive. The idea is to have drugs which "overlap" in their ability to find and kill cancer cells so that none escape. This approach is meeting with some success. Childhood leukemia was once considered a hopeless, invariably fatal malady. Now it is showing a five-year survival rate close to 50 percent.

That work at UCLA, however, involves a completely different approach to leukemia. In those experiments with mice, research scientists found that gamma radiation from a cobalt 60 source, rather than chemicals, can be used to kill all of the mouse's bone marrow cells. Once this is done, new bone marrow—presumably free of malignant cells—can be implanted to begin rebuilding the mouse's blood-making system.

One would be prompted to reply, of course, that such a massive dose of radiation seems like a drastic way to work on leukemia, especially if you're thinking of using it on people. Obviously, much work needs to still be done, but the use of radiation against leukemia seems like a possibility that's worth pursuing. And when the craft of genetic engineering is finally refined, the new bone marrow implanted will probably have been especially designed for each patient.

But such hints of success with leukemia don't mean that doctors are now on the threshold of finding that long-sought-after cure for cancer. Researchers have long suspected that cancer is not a single disease, but a whole family of closely related conditions, all involving loss of control over growth. One property that most cancers exhibit in common is unrestrained growth; the continuous unregulated division and redivision of cells, which rapidly multiply to become expanding tumor tissue. Somehow the genes appear to be closely related to the events of cancer, but what the exact relationship is remains unclear. Similarly, the cell's exterior boundary—the cell membrane—is also somehow involved in cancer, since it is the part of the cell that normally contacts other cells or objects and sends warnings to the genes that space is getting tight. Cancer, then, seems somehow related to the loss of "feeling" at the cell surface, or else the genes don't hear—or respond—when the cell bumps up against its neighbors. Instead of shutting down production, they just keep right on reproducing.

Scientists and physicians studying cancer would like to know, too, whether the genes—sitting at the heart of the living cell's control system—are telling the cell to run wild, or whether the genes themselves are receiving scrambled messages from outside —from the cell's environment—and are reacting in response to bad communication.

Whatever the mechanism, the cells that are exhibiting cancerous behavior are known as transformed cells since they have somehow been transformed from the normal state; from being docile, obedient members of the community, into wild, drunken outlaws trying to take over the whole town.

Their wild quality is best illustrated by what scientists call this loss of "contact inhibition," meaning these cells have discarded their ability to detect normal boundaries. Normal, untransformed cells being cultured in a laboratory dish, for example, merely grow in a single layer until they run out of room, until contact with other cells becomes too close. Then they stop. This is contact inhibition. Conversely, the outlaw cancer cells don't stop growing even when things become densely crowded, but keep right on dividing rapidly, producing new cells which begin piling atop each other in a jumbled, uncontrolled, chaotic mass. This, of course, is also what occurs inside the living body while a tumor is growing out of control. The result is extreme overcrowding, with cancer tissue pushing its way through other tissues until eventually the whole organism—be it man, animal or plant—dies. Death is also speeded by the fact that the rapidly growing tumor creates strong demand for supplies, taking nourishment which would normally go to other body tissues.

One interesting insight that has emerged from cancer research is the suggestion that the growing mass of cancer cells—the expanding tumor—is actually trying to build a whole new person, and do it fast. Somehow the cells seem to have reverted—in part, at least—to the fetal stage, so the message the genes are getting, or are telling themselves, is that they're supposed to be building a whole new individual. Thus they're busy, dividing rapidly, similar to the way the first few cells of a new fetus rapidly get to work building up a supply of tissue. Trouble is, cancer cells seem to lose the ability to shut themselves down, to read the signals that normally say "stop."

Since the genetic information seems somehow scrambled, tumor cells are also seldom able to differentiate into specialized tissues. It's as if these cells are continually yelling at themselves: "Let's go, let's go!" but end up going nowhere, running only in circles. Thus the genes seem to be hard at work, but they're working out of control, building something—but using only part of the plans.

What, then, causes cancer? What is it that suddenly tells one small cell to run wild and start building a whole colony of wild ones?

Nobody has that answer, of course, but researchers are finding strong evidence for numerous causes, including infection by viruses, mutations caused by dangerous radiation, and the damage done by chemicals encountered in the environment. Indeed, several kinds, or families, of viruses have been positively shown to cause cancer in birds and animals, but no virus—or viruses—have been found which cause cancer in man—yet. The search for that human cancer virus is intense, but so far it has been unrewarding. There are, of course, a few candidates. Among them, perhaps the most suspect, are those known as the C-type viruses which are already known to produce cancer—primarily leukemia—in cats, dogs, mice, hamsters, rats, birds, and even apes. None has been found, however, which can be tied directly to causing cancer in man.

These are the difficult viruses that posed such huge puzzles for many years for biologists. Until only a few years ago, it was not known how these C-type tumor viruses do their dirty work. The basic problem was that these viruses carry, as their set of genetic instructions, only a single strand of RNA (ribonucleic acid) instead of a double strand of the master genetic molecule, DNA (deoxyribonucleic acid), like most other viruses. For biologists, the task was to discover how these C-type viruses can invade a healthy cell, insert those single strands of genetic material into the nucleus and take control. The better-understood DNA viruses do this by intermingling their own DNA into the cell's codebook and issuing new commands. For the single-strand RNA viruses, however, no way was known, then, for RNA—which is usually a messenger molecule, not a command molecule—to take command of a cell's DNA. Such a trick ran against one of the central rules of biology, which holds that genetic information only flows in one direction, from DNA to RNA to protein.

In the waning years of the 1960s, however, intensive work in biochemistry led to simultaneous discovery by Dr. Howard Temin and Dr. David Baltimore, in different laboratories, that these notorious C-type viruses carry along with them a handy little enzyme which, once the virus has inserted its RNA into a cell, induces the cell's replicative machinery to run backward, to make a DNA copy of the virus's RNA core. Thus what Temin and Baltimore discovered was the key through which the RNA subverts and takes command of the cell's DNA headquarters.

Another interesting aspect of these strange C-type viruses is that they appear to be transmitted in two ways. First, of course, is through direct contact, the way most of the contagious diseases are spread by exchanging viruses or bacteria among susceptible hosts. Second, and perhaps more important in studying cancer viruses, is the method known as vertical transmission. This means that the virus's genetic material—or at least the genetic information for building the virus—is transmitted from one individual to another through the genes, from parent to offspring. In this case, the viral information is always hidden away deep in the host's genes, and under normal circumstances it doesn't need to come out. It doesn't need to be expressed since it is automatically spread each time the cells divide. For the virus, this would seem an ideal means of reproduction, since there is no need to leave the secure environment inside the cell; and there is no requirement for the virus to gather, process and expend energy for its own reproduction. The strange thing is that these viruses occasionally emerge from this secure spot inside the cell. When this occurs, the particles—the live viruses—can apparently be transmitted normally, horizontally, or for some reason they may also turn their host cell into a wild cancer cell. This suggests that in a case of cancer the viral genes, hidden away inside the cell, may have been activated for some reason, but some defect may have kept the virus from emerging. Thus it may be defective expression of viral information that scrambles the cell's internal signal system to cause cancer.

These C-type viruses, then, exist in rather sharp contrast to the more familiar DNA viruses, some of which are also known to cause cancer in animals. DNA viruses apparently spread horizontally, from one animal to another, and also seem to be handled more easily, recognized as foreigners by the body's normal defense mechanisms against infection.

At present, several DNA viruses are known to be strongly implicated in some types of human cancer. Perhaps first among these is the herpes simplex virus, which is the invisible pest that also causes cold sores on the lips, shingles, and other uncomfortable ailments. Lately, too, evidence has been found that herpes simplex may also be spread between men and women during sexual intercourse, and the implication now is that herpes simplex virus is somehow involved with causing cancer of the cervix in women. Thus some doctors are beginning to view this form of cancer as something akin to a venereal disease, with herpes simplex virus involved somehow.

Studies by epidemiologists indicate, too, that younger women—those passing through their early teens who have frequent intercourse with a number of different men—appear to be particularly vulnerable to this form of cancer.

A second example of a DNA virus that may be causing cancer is found associated with a malady known as Burkitt's lymphoma, a tumor that usually strikes its victims' jaws, producing a huge, grotesque growth that deforms one side of the face. This ailment is currently limited mostly to tropical Africa. The suspect virus in this instance is called the Epstein-Barr virus. There is no proof, yet, that the Epstein-Barr virus is the cause of Burkitt's lymphoma, but the signs are strong, and there is also some indication that the virus could be spread by mosquitos.

Viruses, then, would seem to have a central—but yet undefined —role in cancer. They aren't alone, however. Numerous chemicals are also known, well known, to cause cancer.

Factory workers often show above-average incidence of certain tumors, depending on what substances they may be working with. Shipyard workers often have mesothelioma, a cancer of the abdomen. Studies conducted both in England and the United States show that this type of malignancy is clearly related to long-term exposure to asbestos particles, which are fibrous mineral compounds used extensively during World War II for insulation in warships. Workers in other industries using large amounts of asbestos have also been hit with cancer and, if they smoke, most frequently it's lung cancer. The interesting thing about mesothelioma, too, is that it waits some twenty years after exposure to asbestos before showing up in its victims.

Chemical workers, too, face hazards that only recently have been recognized as causes of cancer. The big rubber firm, B. F. Goodrich, for example, was first to learn the sad lesson that one of the key ingredients in modern plastics—vinyl chloride—causes hepatoma, a disastrous cancer of the liver after long exposure to small amounts. Since discovery of this chemical-cancer relationship, rigid rules have been established to limit the amount of exposure of workers. Some industries have been arguing that the rules are too stiff, or even that the exposure limits are unworkable, but it seems the new rules are going to be enforced anyway.

Inhalation of such chemicals, of course, isn't the only avenue cancer-causing substances can use for entering the body. Ingestion is also considered important, and it has been found that diet plays some role in turning various kinds of normal cells into renegades.

The epidemiology of cancer—the study of what types of cancer occur where, in whom, and when—also strongly suggests that diet is significant. This seems to stand out clearly once scientists look closely at the types of cancer which occur most prominently in various countries, then take a look at the foods these people eat, or at other environmental factors. The types of cancer which occur in Japan, for instance, differ significantly from those occurring in America, in Europe or Africa. The Japanese, indeed, along with Koreans, suffer a much higher rate of stomach cancer than Americans; yet the Americans are known for an increasingly high incidence of lung cancer and bowel cancer. The Asian experience with stomach cancer apparently stems from common use of foods like fermented soya cake, soya sauce and numerous other spicy substances. America's lung cancer rate seems to be clearly tied to smoking, while cancer of the colon has been blamed—perhaps erroneously—on Americans' small dietary use of fiber roughage. An important point, too, is that people who move from one nation to another, and who adopt the new nation's food habits, also seem to adopt the type of cancer that goes with it.

Race also appears to play an important role in cancer. This would further indicate there is a genetic basis, or at least a genetic influence, in the origin of tumors. Caucasians, for example, suffer a relatively high rate of skin cancer, including the deadly variety known as malignant melanoma. To the contrary, blacks seldom get skin cancer, and especially not melanoma. Indeed, reports indicate that one of the few forms of skin cancer in blacks is among certain native Africans who, going barefoot all their lives, sometimes end up with a strange form of skin cancer on their feet.

Despite the attention cancer gets, most people *don't* get cancer. This suggests strongly that some natural defenses against malignancy are built into the body. Indeed, doctors have learned that the normal, healthy body maintains an active, very powerful system that is designed to defend against tumors, against renegade cells and tissues, whatever their source. This mechanism is known as the cellular immune system, which researchers are discovering is a complex, powerful surveillance organization which constantly inspects all parts of the body for maverick cancer cells.

Perhaps the best illustration of the power of this immune system is its ability to quickly poison and reject a transplanted organ. This says to experimenters that the cellular immune system is also constantly searching out and destroying individual cells that are transformed, cells that have gone astray—perhaps in response to a virus, radiation or chemical damage. Whatever cells or substances this system recognizes as different, as "foreign," from the rest of the cells recognized as "self," are immediately attacked. Clearly, the cellular immune system's job is to find, isolate and then kill invading cells or home-grown cancer cells long before they can develop into tumors, long before anyone might suspect something could be wrong.

What this suggests, too, is that cancer may be as much a disease caused by a weakened immune system as it is a result of invasion by a virus, or of chemical damage or radiation. Thus one of the strongest efforts being pursued by cancer researchers is to find some way to beef up a tired immune system, some way to turn it on full force and enhance its power to conquer cancer. They would certainly also like to find ways to turn this immune system off selectively, too, to make the transplantation of organs easier and more reliable. A strong clue that such control of the immune response may be possible someday comes from the observation that patients who've undergone transplant operations—requiring that the immune system be rather crudely disabled—often develop tumors, probably as a side effect of the transplant operation. This would suggest that once the immune system is shut off, put out of commission, maverick transformed cells—which would normally have quickly been found and annihilated—are given a chance to survive and grow into tumors.

What should be remembered, too, is that most of this current work on cancer really amounts to cellular engineering rather than being true genetic engineering. Indeed, scientists are largely convinced that cellular engineering will become a successful enterprise much sooner than genetic engineering will, mainly because of the large information gaps about genetic processes. In some instances science hasn't reached the point yet of even knowing which questions to ask. Nonetheless, when the day comes when scientists do begin manipulating genes in human patients, one of the prime targets will be to find ways to strengthen that important immune system. Doctors will be interested in building this system into a stronger, more versatile weapon, and they'll probably find ways to make it more controllable.

While this work on disease may be part of the promise of genetic engineering, for many people the mere mention of genetic manipulation creates unpleasant visions of babies being produced in test tubes, later to be moved into coldly mechanical wombs when well on the road to "birth." Recent developments in biology and medicine mean this is becoming ever more likely. Actually it might represent a real boon, a true convenience for a woman faced with a difficult, painful pregnancy during a long, hot summer. But it does raise serious questions as well.

The question that will most need asking, of course, is whether there is any real advantage to growing babies up mechanically, even if they are perfect, in a world already loaded down with people who are in reasonably good genetic health. On the other hand, if demonstrably better babies result from such strange new technology, then such a child-producing system might come into wider use than many people now suspect is even possible. Nonetheless, if it turns out to be merely a way for producing more people faster, who needs it?

At first glance, certainly, research into the problems of fertility—or rather the problems of infertility—seems pointless in a crowded world. Still, it's not pointless to couples intent on having their own children but who don't succeed because of medical problems. Since this is apparently a need that is felt very strongly by many people, it represents one area of medical research that is going to keep expanding. Services already available include artificial insemination—with some 10,000 inseminations now done annually in the United States alone—and sperm banks in which live sperm can be quick-frozen and kept viable for years. Too, there have been well-publicized efforts—notably in England—to implant an egg-sperm combination, a developing early embryo, into a woman's uterus. This has apparently been done successfully, with the patients' identities kept strictly secret, but the reports that children produced by this approach were normal in all respects were not substantiated.

One should ask then, when speaking of genetic engineering and normalcy, what can really be considered normal? Is it genetically normal for a large percentage of American men to be at risk for heart attack? Is it normal that more and more people—especially Americans—become victims of emphysema, rendering them barely able to function without gasping for every breath? Are these diseases, these frailties, genetic in origin? Or can we blame the fumes from factories, our diets, or something else? The answer, probably, is something like "all of the above."

When it comes to the problem of heart attacks, however, many doctors are concluding that American men—as well as the men living and working in other westernized nations—are probably digging their own graves with their teeth. Diets laden with rich animal fats have long been recognized as a probable hazard, especially for some men. Still, diet doesn't seem to tell the whole story. Despite the fat-rich diet popular in America, there seems to also be a strong hereditary influence in heart disease. Basically, it can be said with some confidence, that if a man has healthy, long-lived parents, chances are he won't suffer a massive heart attack and die in middle age. If, on the other hand, he's born into a family in which the men almost routinely succumb to heart attacks early, chances are also strong he'll fit snugly into that pattern however well he watches his diet and exercise. Heart disease, then, would seem to be rather well programmed into the genes—at least to some degree—so that all an individual can do is try to make the best of what he's got, paying close attention to family history, watching his diet, and exercising carefully and regularly.

In the future, certainly, genetic engineers will also be wanting to help combat heart disease. The key approach via genetics will probably lie in finding out which gene—or set of genes—is involved in heart disease, then find some way to replace this faulty genetic code with healthier chemical information. Obviously the techniques for accomplishing this still don't exist, and probably won't for a long time yet, but the possibilities are still interesting—and they will be explored.

As mentioned earlier, another approach to such genetic problems may be to snip out a little of the person's tissue—actually take a sample—very early in his life. This sample could then be frozen, if that technique, too, is perfected, for storage. It could later be thawed and be grown up into a healthy new organ such as a heart for use when the original gives out. What this will require, first, is learning what genetic or biochemical signals are needed to tell the tissue sample: "Make me a new heart." That this may become possible is already evident, since scientists have found that by applying the right substances they can induce cells taken from the liver to produce proteins normally made only by brain tissue. This demonstrates that even very specialized cells might be tricked into doing new jobs if one knows the right chemical codes and how to apply them.

Similar techniques might also find use against that lung-destroying malady known as emphysema, especially since some cases of this disease have been related to a genetically based enzyme deficiency. One could certainly hope that this deficiency might someday be spotted early in a person's life while he's still young enough to be helped. According to the work already done by a research team at the City of Hope National Medical Center in Duarte, California, a significant number of today's emphysema victims suffer this deterioration in lung function because of an inborn shortage of a biochemical called alpha-anti-trypsin. When this protein is in short supply, courtesy of a genetic mistake, the victim's lungs are slowly destroyed—actually digested—by chemicals the body normally uses for scavenging foreign substances that find their way into the lungs. It is suggested by researchers that when too little alpha-anti-trypsin is available, normal lung-cleaning chemicals remain active too long and begin digesting lung tissue, too.

Before long the lungs lose much of their elasticity, their stretchability, as the tiny air sacs—the alveoli—become deformed. At the same time, there is also a progressive loss of the lung tissue's ability to exchange waste carbon dioxide for fresh new oxygen. Normally the symptoms of emphysema slowly grow worse as the architecture of the lungs is destroyed.

Fortunately the researchers at the City of Hope—which was once an important tuberculosis hospital serving southern California—also discovered that one of the best ways to work against this ailment—alpha-anti-trypsin deficiency—is to keep the patient from smoking. They found that the combination of tobacco smoke and the genetic abnormality lead to rapid deterioration of lung tissue, accelerating the progress of emphysema.

Future genetic treatment for emphysema will probably involve the same approaches that will be used for people with heart problems, such as growing up a whole new set of lungs in tissue culture, or even the implantation of healthy new genes to correct the enzyme deficiency.

With all these possibilities, all these signs of progress and all the remaining challenges, where, then, does the world of medical science stand today?

Basically, medicine today is poised now right on the threshold, about to step gingerly into a whole new world of treatment, ethics, opportunities and, alas, controls. Research will continue, doctors will still be colliding with disappointing dead ends. But there will also be many successes as, one by one, these genetic diseases are finally cracked. Many of them will first simply be eliminated by widespread abortion. Much later, gene therapy will come into use, and genes will actually be swapped around for the cure of specific genetic ailments in specific people.

The investment in time, talent, and medical resources will continue to be large, and this whole research and development process is going to keep raising important questions. Most will involve that old nagging problem of whether our resources are being well spent curing these difficult, relatively rare diseases when we could be spending them on people we already know how to cure.

Someday, someone is going to demand an answer.

4

More Food for a Hungry World

GARDENERS WHO'VE always counted green thumbs necessary for success in the soil will be rethinking that old idea once genetic engineers come to grips with important food plants. Talent— especially in agriculture—will always be in demand, but the teams of biologists, botanists, geneticists, and biochemists now beginning to rearrange the genes of the world's most vital food crops will soon be finding ways to make even the brownest, driest thumbs green up on demand.

These promised developments in the world of agriculture should not be dismissed as improbable, nor should they be taken lightly. In this one vital area—the world's food supply—genetic engineering probably has more to contribute than any other branch of science or technology. When life scientists finally have the means and methods worked out, the changes in food production—in terms of both quantity and quality—will be truly dramatic. Hunger may never be completely banished, but genetically engineered food crops are going to provide one great leap in that direction.

The expected developments in food science, indeed, sound a little like science fiction, but the possibilities are certainly real. Here are a few of them:

• Growth of exotic new species of trees, fruits and vegetables— and especially grains—produced through genetic combinations never before possible. This will be achieved by "mating the unmatable," crossing peas, for example, with cantaloupes. What the results will be nobody knows yet, but bets are that the changes will be startling and probably productive.

• Transfer, finally, of the famous nitrogen-fixing "nif" gene—or genes—from the leguminous plants like peas and soybeans into other food plants like wheat, corn and rice. The goal is to make these grain plants gather their own fertilizer from the air. When this is achieved, the impact on world food supplies will be profound.

• Rebuilding of the metabolic systems of important food crops like soybeans, making them more efficient, more productive through better use of sunlight, the solar energy collected by their leaves.

This brief list represents only a few of the possibilities that will come from genetic engineering, but they do suggest what goals are being pursued as scientists begin rearranging and recombining the genetic material from a wide variety of plants. It should be obvious, of course, that the advances in agriculture will follow on the heels of the elegant, important work that has already been done, and is still being done, toward producing better crops through conventional methods of crossbreeding. Crossbreeding—which hasn't involved genetic engineering at all—has already given the world several vigorous new types of corn, new wheats, better rice and improved barley, oats, and rye, and even one new type of grain: a cross between wheat and rye known as triticale. This work is what led to the famous Green Revolution which, though much maligned of late, still remains a monumental achievement in agricultural science.

Despite severe criticism of the Green Revolution—based mainly on the fact that the new super-wheats and super-rices require large amounts of expensive chemical fertilizers—it can still be considered a triumph, or really a series of triumphs in the art of plant breeding. Even though the productive new grain varieties introduced into the Far East and elsewhere haven't produced all the impact hoped for, other new strains of wheat are just now being introduced to American farmers, and the prospect is that they can boost yields in the United States by 15 or 20 percent. Major American grain companies have been working to create and introduce these new kinds of wheat for years, and they're now becoming available for commercial planting.

Much of the work that led to production of new wheat varieties —and especially those types involved in the Green Revolution— was done through the International Wheat Improvement Project in Mexico under Dr. Norman Borlaug, funded by both the Rockefeller and Ford foundations. The similar work that produced new and vigorous rice varieties was done at the International Rice Research Institute in the Philippines, also supported by the two American foundations.

Basically, the new wheat varieties were bred specifically to grow vigorously and yield bountifully in areas such as India and Pakistan. Once introduced, they briefly increased yields to the point where some developing countries almost become self-supporting, and a few even became net exporters of food. The same is true for the new rice varieties introduced into the Far East. In both cases, however, the Arab-induced worldwide energy shortage, which greatly boosted the cost of fertilizer, brought progress to a sad and sudden halt.

The new strains of wheat and rice were bred to grow extra full, extra heavy heads of grain that formed atop rather short, strong stalks which helped prevent lodging, or falling over. As a result, the increase in yield—if given ample fertilizer and if well managed —was substantial, especially when compared to the old favorite native varieties.

In addition to the fertilizer problems, in some places the new grains didn't always match well with the foods the people were accustomed to, so new types of wheat and rice weren't always accepted with unvarnished enthusiasm. Nonetheless, further manipulation of these plants through crossbreeding is continuing, and the promise is that the improved supergrains can be made to match people's tastes closely. Until fertilizer prices fall, however, the Green Revolution can be expected to remain stagnant.

With corn, however, it's a different story, especially in the United States, where fertilizer has generally been abundant and where the tools of large-scale farming are put to highly efficient use. Indeed, the increase in corn yields in the past half century

Former University of Chicago president Dr. George Beadle works here with what he believes may be an important genetic resource for the improvement of corn. Beadle argues that this Central American plant, called teosinte, is closely related to the ancient plant from which modern corn arose, with the help of man. To prove that teosinte could have served as a food for primitive Americans, Beadle has grown, harvested, and ground the edible parts of teosinte grains to produce foods.

has been nothing short of phenomenal, and yields have been improving steadily ever since the first hybrid corn varieties were introduced in the early 1920s.

Corn has its own interesting history, a history that began in the Americas—most probably in Central America and Mexico—where plants that are descended from the original wild corn are still found growing wild today. There is some dispute over this, but geneticist Dr. George Beadle, a Nobel laureate and former president of the University of Chicago, suggests that one of the native grasses found in some areas of Central America—a plant now called teosinte (pronounced tee-oh-sin-tee)—is probably the direct, almost unchanged and unselected descendant of the wild grass which once gave rise to what we now call modern corn. A few specialists are still arguing heatedly over whether teosinte is really that closely related to corn and with corn's ancient ancestor; but Beadle's arguments appear to be increasingly persuasive, and he has also proved that the hard teosinte kernels, once free of their tough outer husks, can be used as food.

Scientists arguing against Beadle's interpretation usually point to samples of corn pollen found in archaeological digs near Mexico City. The pollen is dated as far back as 80,000 years, which suggests that corn may have been a separate species, distinct from teosinte, long before human beings arrived in the Americas to begin the selection process.

On the other hand, Beadle persists in his argument that teosinte is today's representative of ancient corn, and he noted recently:

> The linking of corn to teosinte may have important implications for corn breeders. It could provide them with a new reservoir of genetic diversity to meet their particular needs. The relationship of corn to teosinte has been debated for many years, but it is a relationship we can't afford to overlook. For thirty-five years attention was diverted from teosinte as a plausible source of genetic materials for breeders, and this—I fear—was a disservice.

Regardless of this continuing argument among specialists, there is now little question that today's corn—which can be seen on summer days rippling and swaying on vast fields, in large gardens

A single "ear" of teosinte, shown in the hands of Dr. George Beadle of the University of Chicago, illustrates how small the grain-bearing structure of this plant is. Compared to the rich, full ears of corn produced on modern hybrid corncobs, teosinte would provide a meager meal indeed. Beadle nonetheless argues that teosinte is the closest living relative— probably a direct descendant—of the wild Central American grass which eventually became modern corn.

and even in tiny plots—is an important and enduring gift directly from the Indians of America. Over uncounted centuries these original Americans took this once-wild grass and gradually maneuvered it toward aristocratic rank among the world's wide variety of food crops. Even though the Indians lacked the skills and tools of modern genetics, they managed—with the unaided eye, bare hands, and some imagination—to repeatedly select out the more desirable mutant varieties of corn and finally domesticate this crop. This job was so thoroughly done, in fact, that today's ultra-productive varieties of corn have become so completely domesticated that they are biologically helpless, now needing the nurture and protection of man to survive. Obviously, domestication has its price.

It was corn that sustained what was left of the small band of Pilgrims at Plymouth after their wheat crops failed. According to the old story, Indians taught these new settlers to plant, cultivate, and harvest this strange new grain which was native to the New World. Soon, as the settlers began expanding their territory, moving inexorably westward, corn went along with them to become an increasingly important part of the diet. Today, of course, corn is a staple food and is also the basis for a huge, productive livestock-feeding industry.

Some historians might argue that corn eventually became the grain that built the Western Hemisphere, since this crop's ability to produce high yields, even in ancient times, is what gave the early inhabitants of the Americas—the Incas, the Mayans, and Aztecs—enough food security so empire-building could begin.

Corn, too, became equally important in other parts of the world. Once the use of corn had been introduced to Europeans, it took only a few years for this valuable food to find its way—via Portuguese sailors—into Africa, and also with Columbus into Europe. Indeed, within two generations of the discovery of the Americas, corn had been spread through most of Africa and into parts of India, China and Tibet.

Still, the real modernization of corn is another more modern story. As a result of the first sophisticated and complex corn-breeding programs that began in the 1920s, the production of corn

Cobless corn. Corn breeders looking for ways to increase yields by 250 to 300 percent are experimenting with this type of tassel-seed corn. The trick they're trying to perform is to move the corn plant's grain production site from the ear to the tassel. Commercial use of such cobless corn still seems a long way off, but this work serves as one good example of what is being attempted through normal crossbreeding techniques. Even more dramatic results are thought to be possible someday through genetic engineering. This type of tassel-seed corn is being developed by DEKALB AgResearch Inc.

today reaches almost overwhelming proportions. Because of these newer breeding efforts, there is now more real foodstuff inside a single grain of some modern races of corn than was found in a whole ear of prehistoric wild corn.

This great boost in productivity—both from the standpoint of larger ears and more vigorous corn plants—began with the deliberate hybridization that involved the crossbreeding of different types of corn, types known to be carrying desirable genetic characteristics for vigorous growth, for good grain production, and for disease resistance. Corn varieties blessed with these traits began emerging first from the laboratories run by G. H. Shull at the Carnegie Institution, by E. M. East at Harvard University, and by D. F. Jones at the Connecticut Agricultural Experiment Station. These profitable, productive new varieties were quickly adopted by farmers all across America, and by 1969 some 99 percent—or more—of the huge U.S. corn crop came from hybrid corn. Today, with most of America's corn going for use as animal feed, one efficient farmer working in the U.S. Corn Belt can, with the help of only one man, grow enough corn to produce meat and other livestock products to feed from 300 to 400 people. Indeed, yields have increased from the 1929 average of 25.7 bushels per acre to an average now of more than 80 bushels per acre; and the increase is continuing, albeit more slowly.

But most hybrid corns don't normally breed true. Thus a farmer can't expect to save part of one year's crop to be replanted a year later and still hope to grow the same quality or type of corn. New "seed corn" must be grown every year, and this has led to the formation and growth of large seed companies, firms such as DeKalb and Cargill which use large tracts of land for growing up reliable new seed for each year's crop. This may sound complicated and expensive, but it has—when combined with the use of high-nitrogen fertilizer, mechanical farming, and new techniques —made the United States Corn Belt the most productive agricultural civilization the world has ever known.

There is much more to the story of corn than just increases in productivity, dramatic as they are. The most recent development in corn breeding involves the production of new corn varieties known as high-lysine—or opaque—corn which in time promises to significantly improve levels of nutrition in the heavily corn-dependent areas like Latin America and Africa. Such improvement of nutrition is possible because these new corn varieties—named "opaque" because of the nontranslucent quality of individual kernels—contain increased amounts of the essential amino acid lysine. As a result, this new corn is almost as valuable for the human diet as whole milk. Thus, in regions where diets are based mostly on corn, these new opaque varieties are providing a dramatically effective, reasonably simple way to alleviate chronic malnutrition.

Again, however, as was the case with the other new grains, opaque corn has some differences in color, texture, and milling qualities which make it unfamiliar and thus difficult to introduce into a population's diet, however nutritious it may be. Because of such problems, additional work is under way toward matching even newer varieties of lysine-rich corn to local diets while still retaining that important boost in nutritional value. Success doesn't seem too far off, and in the meantime efforts are still being made to induce more people to accept use of the present lysine-rich corn varieties.

What should be realized, nonetheless, despite these nearly miraculous improvements in the world's favorite food crops through conventional crossbreeding, is that scientists are facing the prospect of eventually running out of ways to improve crops further with these techniques.

Already, long before they've begun to run out of useful genetic resources, there is genuine cause for concern that, as more and more of these new superproductive crops are planted, even in remote areas of the world, many of the older varieties and even some wild grains will be crowded out and lost. This is important because older plants and wild plants probably contain genes which will be useful in future crossbreeding experiments, coding for characteristics like disease resistance, dwarfism, or resistance to drought. To help avoid this problem, some plant research centers are now beginning to set up seed banks where gentically important types of seeds can be stored and thus perpetuate these older varieties.

One good example of the effort going into such preservation of genetic resources is the work of the United Nations Food and Agriculture Organization (FAO), which has set up a special Genetic Resources Unit inside Turkey's Aegean Regional Agricultural Research Institute in Izmir. In 1964 this was the first such unit specifically established to collect, store, and study the fast-vanishing genetic resources found in both the cultivated and the wild plants in the Middle East. Among these valuable plants are primitive relatives of some of the world's most important crops— wheat and barley, for instance—plus some legumes which produce important forage.

It's unfortunate, perhaps, that the Turkish government also chose to use this seed storage unit for preserving opium poppy seeds during the brief period when opium cultivation was being actively discouraged—at the urging of the U.S. government.

The idea behind creation of such "gene banks," of course, isn't really anything new. Plant breeders have been collecting, sorting, and storing plant materials gathered from all over the world for years, especially for use in plant-breeding experiments. Indeed, large reference collections have also been established in the United States, with the U.S. National Seed Storage Laboratory at Fort Collins, Colorado, one of the most important and extensive facilities. In Russia, too, valuable collections exist which were established by N. E. Vavilov and his coworkers in the 1920s and 1930s in centers such as Leningrad. It's fortunate now that these seed collections weren't simply disposed of, tossed out, during the sad, wasteful and disastrous years when T. D. Lysenko misruled Soviet agriculture.

Still, this newer seed-storage laboratory established by FAO in the Middle East has its own special importance, because the lands reaching from Turkey to Pakistan—including parts of Israel and Lebanon—represent what might well be called the cradle of Western agriculture. Wheat and barley were probably cultivated in the Middle East as much as 10,000 years ago, and modern archaelogical detective work hints that a system of mixed agriculture was probably well developed by the time of the eighth century B.C. Samples of barley and several kinds of wheat—relatives of the various wheats in wide use today—have been discovered in some archaeological digs, and they may date as far back as 6,000 or 7,000 B.C.

At present the biggest problem—other than persistent warfare —now facing agricultural scientists working in the Middle East to preserve old species of plants, is that the increasing use of newer, more productive varieties of grain means improved varieties are rapidly replacing the older local stocks, save for a few remnants grown in a declining number of remote areas. The truly wild descendants of ancient types of wheat and barley can also still be found in some localities in the Middle East, fortunately in a few areas where they are now protected. Wild barley is still found growing amid the ruins at Ephesus.

In addition to seed storage, other methods of genetic preservation are being worked out. One of the most promising—but still unused—ideas is that plant tissues can be frozen alive for long-term storage. Seeds, too, have been frozen, and some later experiments with the freeze-drying of seeds indicate that seeds put through this process actually germinate better and more consistently than those stored in seed banks under conventional conditions.

Research has shown recently, indeed, that plant tissues snipped from the growing tips of adult plants can be quick-frozen—down to $-196°$ Celsius—and can be stored at this ultra-low temperature for rather long periods. Later these tissues can be carefully thawed, nurtured in growth medium and induced to grow into fully developed, viable adult plants. The new plants grown from previously frozen tissue samples are genetically identical to the plants from which the tissues were snipped. As Dr. William Nelson at the GTE Laboratories in Waltham, Massachusetts, explained it: "At present, people trying to preserve genetic stocks have to do it with seed storage. But with seeds you don't get one hundred percent transmission of the genetic materials. With our [freezing] method, with tissue culture, you do get identical copies."

As Nelson mentioned, stored seeds often carry the combined

genetic traits from two parent plants, in most cases; thus, many seeds won't necessarily produce an exact copy of either of the parents. In other words, every time a seed is made—except when a plant is able to "self," depositing its own pollen on its own pistil —the genes get reshuffled into new combinations. This is especially true in the wild, where constant cross-pollination is likely to be more important. Thus, in tissue culture, where no seeds and no breeding are necessary, truly exact reproduction of one plant is assured since no shuffling of the genes occurs.

Of course, gardeners and fruit growers have long been taking advantage of this fact simply by using cuttings or grafted plant material to perpetuate valued types of flowers and fruit. More recently, however, botanists discovered that if they isolate the new cells from the growing tip of a plant, they can come up with tissues that are not contaminated with viruses. In nature, unfortunately, most plants are known to serve as hosts for numerous kinds of viruses. It turns out, though, that these tiny, invisible creatures appear to inhabit only the more mature parts of the plant. Tips snipped from the growing end of a branch or stem are thus usually virus-free.

The interesting point is that through use of this knowledge botanists have been able to take an immature piece of tissue, or even a single cell, then culture it into an adult plant that is itself free of virus infection. The result has been introduction of vigorous new generations of plants which are more productive, bloom bigger and brighter, and outdo their virus-laden relatives completely. In the propagation of geraniums, for example, use of new virus-free tissue has led to startling improvements in the size of flowers, in the brilliance of color and the vigor of the plants.

It's easy to see, too, why plant scientists are excited about the prospects of freezing tissue samples when the results can be so good. As noted, the group of researchers at the GTE laboratories purposely chose to freeze only the tips of plants—in this case carnation plants—in their experiments. Since they're using only tips, it means they're putting tissues into cold storage which are virus-free, and the thawed tissue will thus be free of infection when it is grown up into an adult plant.

Michael Siebert, another researcher at the GTE Laboratories, explained that during these tissue-freezing experiments, the carnation cells were placed in a special solution, after which liquid nitrogen—at –196° Celsius—was poured into the vial containing the plant sample. Next, the whole vial was dipped into a bath of liquid nitrogen. After this treatment, the live plant cells were kept frozen for at least two months, then thawed.

The results, he added, "demonstrate that a shoot apex [the tip] can be frozen . . . and successfully thawed in a viable state. Furthermore, the fact that shoots and, subsequently, plants, can be obtained from surviving apices [tips] means that the morphogenic potential [the ability to grow into an adult plant] is not necessarily destroyed by the freezing process."

Finally, Siebert said, with refinement of this new freeze-thaw process it may become possible to establish plant-tissue banks for the preservation of important botanic genetic combinations. This would avoid the problems associated with reshuffling the genes during seed production.

This quick-freeze idea isn't really new. Plant scientists and food researchers have been experimenting with methods for freezing plant materials for decades as an effective method for preserving foods. It was learned very early, for instance, that normal, slow freezing tends to damage the plant tissues, actually damaging individual cells by allowing ice crystals to form inside. The crystals bear sharp edges which break up the delicate membranes inside the cell and the larger membrane surrounding the cell. As a result, slow freezing tends to destroy the taste and texture of frozen foods, so the industry was finally led toward development of quick-freeze techniques.

Outside the food industry, too, scientists—and especially doctors —have met with considerable success recently in use of the quick-freeze. The best examples currently for use of ultralow temperatures for storage of living organisms or tissues comes from the increasing use of frozen blood components and from the establishment of "sperm bank" services, in which a man's sperm cells can now routinely be frozen and stored. This service is aimed, of course, at men who—having had vasectomies—want to keep some viable sperm around "just in case."

There are, as one might suspect, some problems involved too. One case, in California's San Francisco Bay area, involved a man who found that the commercial sperm bank in which he had stored his "seed" had let all samples thaw and die. At last word he was suing for $5 million.

Despite such interesting—but minor—problems, it should be obvious that all kinds of storage techniques are going to be increasingly important in the future. In addition to the problem of newer food crops pushing older varieties aside, agricultural specialists also point with some apprehension to the growing tendency for single cropping. Too much dependence on one crop—especially a genetically uniform crop—can lead to disastrous food losses such as the potato famine that hit Ireland in the 1840s, or more recently the rapid spread of a new strain of southern corn blight through much of the U.S. Corn Belt in the early 1970s. The loss of some 15 percent of the nation's corn crop in one year—1970 —is a good example and a powerful warning of what is possible when major crops become too uniform genetically. In the case of the corn blight, a single variety of very productive corn—carrying a special gene for male sterility—was widely planted through much of America's rich corn-growing country. All it took were some slightly abnormal weather conditions, the presence of a damaging, rapidly spreading disease organism, and a genetically vulnerable crop to set up the conditions necessary for near-disaster. Our losses were substantial.

Since then, however, there has been a near-frantic scramble to introduce substitute types of corn, changing the genetic base of the crop enough to eliminate that susceptibility to blight. Nonetheless, some specialists still warn that American crops are too genetically uniform, that little has been learned through the costly lesson taught by corn blight. The lesson is that no nation should depend on a single crop for its food—or for its entire livelihood, for that matter—because all it takes is a minor genetic flaw, or bad weather or a new type of disease, to tip the balance the wrong way. Indeed, at a time when American granaries were overflowing with surpluses such a gamble might have seemed worth taking, but now, when food is in short supply worldwide, we can't afford to use a genetically narrow base for a major food crop.

Unfortunately, the time when American food production will finally get back to that comfortable condition of surplus can't be forecast, but it may not be too far off if genetic engineering specialists can perform one of the neatest agricultural tricks ever: making plants like corn, wheat and rice produce their own nitrogen-rich fertilizer from the air.

Nitrogen, the most abundant constituent in our air makes up almost 80 percent of the earth's atmosphere, but it is basically of no use to man in its gaseous state. Nitrogen must first be converted into substances man can eat; the amino acids which make up the proteins found in corn, wheat, beef, eggs, fish and other foods. This means, then—if scientists ever hope to get a firmer grip on food production—they've got to find ways to make more nitrogen available, especially in the soil, for food crops. This is done now by dumping nitrogen-loaded fertilizers directly onto the fields, as is presently done in agricultural areas. Given time, however, it may be done by changing or enhancing the natural biochemical process known as nitrogen fixation. Research in this latter direction—toward better, more efficient nitrogen fixation—is now being supported by the U.S. National Science Foundation at research centers like Oregon State University, several campuses of the University of California, Stanford University, Purdue University, the University of Wisconsin, and Northwestern University.

Basically, nitrogen is processed through the earth's atmosphere in a definite cycle—quite naturally known as the nitrogen cycle— which begins with the gaseous molecules of nitrogen in the air. Some nitrogen is taken directly from the air and is processed in the tissues of living organisms. Later in this cycle, when living things die and decay, the nitrogen compounds also break down and their store of nitrogen is either moved into the soil or back into the atmosphere.

Even though decay processes put substantial amounts of nitrogen into the soil, most of the nitrogen found there arrives by another process, nitrogen fixation. This is a task accomplished by bacteria that live in association with the roots of certain plants, especially the legumes. Additional nitrogen is also "fixed" in the soil by some free-living bacteria, and more nitrogen enters the ground directly from the atmosphere, produced or "fixed" by the flashing of lightning. At the top of the atmosphere, too, ultra-violet light from the sun also creates—or fixes—some nitrogen that can be useful for living organisms.

Some chemists have also suggested, rather convincingly, that these latter processes—lightning and ultraviolet light—could have been the energy sources which caused the original amino acids to form in the earth's primitive atmosphere or oceans, leading eventually toward the creation of life. It is also suspected that the formation of these important nitrogen-laden amino acids—which are often called the building blocks of proteins—was the first crucial step toward the formation of living things, that it may account for the origin of life some 3 billion years ago when molecules of nitrogen and hydrogen were combined in the atmosphere. This scenario suggests that they first formed ammonia, which was then washed into the sea to mix with other organic chemicals to combine into even more complex compounds. Eventually, over eons of time, these amino acids were mixed and matched, broken and dispersed, and built into all the different combinations, one of which finally gave life its start.

Even though these two forms of nitrogen-fixing energy were perhaps of crucial importance in getting life started, today the most important method for fixing this nutrient is through bacteria living with, or inside, the roots of legumes. Some specialists have estimated that 100 million to 200 million tons of ammonia are produced annually worldwide through this symbiotic plant-bacteria relationship. For purposes of comparison, the human agricultural enterprise—the modern processes for building chemical fertilizers rich in nitrogen for farming—now manages to produce some 30 million tons of ammonia annually. Industrial production is expected to increase dramatically, however, to 100 million tons of fertilizer per year by the end of this century.

Even in ancient times, of course, farmers knew that some types of plants, now called the legumes, for some reason make the soil richer after they've occupied a field for a season or more. Indeed, one of the obvious benefits of regular crop rotation is enrichment of the soil with the nitrogen that's left over from a leguminous crop; nitrogen that was stored in the roots of plants like alfalfa, beans and peas.

The legumes have this ability—this talent for taking nitrogen from the air and turning it into ammonia—because sometime in the deep, dark past of evolution they formed a close, cooperative relationship with a family of bacteria called the *Rhizobia*. This symbiotic relationship requires that the bacteria invade the roots of the leguminous plant and that the plant, in response, grow some little roundish root nodules or lumps in which the bacteria find a comfortable, almost oxygen-free home. In this relationship, it's up to the plant to supply ample energy for the bacteria, which in turn do their job of converting nitrogen from the air into forms usable by the plant.

What scientists find of interest here, in addition to the production of fertilizer, is that the bacterium is able, given the environment of the legume root, to weld together the chemicals that make up ammonia. The bacterium does this at low temperatures and at low pressure while man, trying to do the same thing, must build huge chemical factories that require great amounts of energy, high temperatures and elevated pressures. The most successful industrial method, so far, is the Haber-Bosch process, which forces molecular nitrogen to combine with hydrogen at temperatures up to 500° Celsius and at pressures 200 to 300 times higher than atmospheric pressure.

It should also be noted that other bacteria besides *Rhizobia* exist which can similarly fix nitrogen, but they seem to be able to do it without the need for the shelter of a root nodule. These, known as the free-living nitrogen-fixing bacteria, include *Azobacter* and *Klebsiella*, plus some of the blue-green algae, *Anabaena*. These bacteria can be found living inside animal intestines, while some nitrogen-fixing fungi, the antinomycetes, are found living in symbiosis with algae to make the lichens. Still the most important, especially in terms of bulk nitrogen fixation, are the *Rhizobia,* which live inside those cozy root nodules.

Ideally, then, because of the expense involved in production of commercial fertilizers, many scientists would like to find some way to duplicate that natural process of nitrogen fixation, but still do it in large vats on an industrial scale. What this would involve is finding—or making—the proper bacteria, then finding the right combinations of light, temperature, and chemical balance, plus nutrients, to induce them to roll out ammonia by the ton. With genetic engineering, too, it may become possible to find ways to improve the capabilities of bacteria, actually redesigning them to take on this task.

Other scientists—especially the botanists, biologists and bio-chemists—however, are looking at a different approach, hoping to find a way to make *Rhizobia* or some other nitrogen-fixing orga-nism form a symbiotic relationship with plants other than the legumes such as corn, wheat, rye, or rice.

Unfortunately, this will probably require changing the tastes and habits of bacteria, and it will probably also require changing some of the basic interior chemistry of these plants so they'll closely match what the legumes normally offer to support these creatures. If no way is found to establish this close symbiotic rela-tionship, ways may still be found for spurring nitrogen fixation without trying to grow new root nodules. Indeed, there are al-ready some hints that this may be possible coming from several laboratories.

Numerous approaches to the problems of nitrogen fixation seem possible, but scientists have noted that only a few of these appear very promising at the moment. According to several agricultural scientists attending the famous 1975 Asilomar Conference on Re-combinant DNA—the meeting where the first basic ground rules for genetic engineering experiments were laid down—there are now four broad approaches which could someday lead to enhance-ment of different plants' ability to fix their own nitrogen:

• First, to improve the nitrogen-fixing ability of existing leg-umes. The goal would be to find new ways to make these plants

—such as alfalfa, peas, and soybeans—more efficient in their use of energy; to find ways to get them to squeeze more value out of the sunlight they receive.

• Second, it may be possible to exploit several types of bacteria now known to live free in the soil. Some of these small organisms seem to coexist with a number of grasses that grow in tropical areas, giving hope they may be adaptable to the more productive grasses that grow our food, wheat, corn, etc. The goal would be to transfer such bacteria to major crop areas where, if adaptable, they could be encouraged to provide a nitrogen boost for important food crops.

• Third is the prospect of finding or making new mutant strains of nitrogen-fixing bacteria. The goal would be to genetically turn on their machinery for nitrogen fixation; then under the right circumstances make them pump out vast amounts of ammonia on command. One can indeed visualize a system in which such bacteria—well-housed, well-fed, and pampered—could produce large enough amounts of ammonia to have important impact as commercial fertilizers. At present this idea sounds a bit farfetched, but efforts to trick bacteria into producing useful chemicals are already well under way in many countries, probably most vigorously in Japan. Remember, too, that many commercial products are the result of captive organisms, e.g., yeast; and such living organisms are the source of important medicines, e.g., penicillin. The U.S. Army, at its Natick, Massachusetts, research center, has recently found some fungi or molds which produce enzymes that can be used for turning waste paper and other wood fiber products into sugar. This suggests, of course, that more developments in this whole field are possible and may be coming soon. And one of them just might involve a new, genetically engineered way to fix nitrogen cheaply.

• Fourth are the efforts to change basic plant genetics, striving to put the proper genes into wheat, for instance, so—like legumes —it could play host to *Rhizobia* bacteria. If this can be done, some scientists suspect that a number of important crops could even be made self-sufficient in fertilizer—and perhaps even add more nitrogen to the soil than they take out. One problem, however, is that these bacteria require that the plant supply unusually large amounts of energy for the fixation process. Scientists also suspect that some of the crops they'd most like to give nitrogen nodules to might not—in their present form, at least—be able to supply enough extra energy to support the process.

This barrier exists because every known organism that fixes nitrogen exhibits high energy requirements. Indeed, the enzyme the bacteria use for building ammonia is an important biocatalyst known as nitrogenase which, unlike other enzyme systems needing energy, demands unusually large amounts of life's basic energy-carrying molecule, ATP (adenosinetriphosphate) to function. Thus, a plant blessed with the ability to produce its own fertilizer is also a slave, required to also devote a substantial fraction of its energy supply to this task. Thus, as they say, "There ain't no such thing as a free lunch." Pea plants, for example, appear to send something like 30 percent of the total carbon they fix in photosynthesis down to the root nodules. Better than half of that carbon stays there to build and support the nitrogen-fixation machinery. This means, then, that nitrogen-fixation—while it gives a good boost to the plant—is also costly for a plant that is striving with all the energy it has available to get on with the job of seed formation.

An interesting point, discussed by Dr. Ray Valentine, of the University of California at Davis, is that "the high energy cost of the nitrogen-fixation process may even explain why most higher plants have not evolved symbiotic relationships with nitrogen-fixing bacteria. They simply may not be able to afford the energy."

In work with those free-living bacteria—as discussed by Dr. Robert Burris, from the University of Wisconsin at Madison—one group in Brazil, headed by Dr. Johanna Dobereiner, has discovered an interesting family of bacteria which can synthesize ammonia without hiding inside a root nodule. One species, *Spirillum lipoferum,* is unique because it grows inside the root of the plant, yet it has no need to use a big round root nodule.

"Dr. Dobereiner has examined numerous tropical grasses," Burris noted, "and she finds that this association [of plant and bacteria] will give her something like 80 pounds of nitrogen per acre per year, which is very good fixation. The difficulty so far lies in putting it to work in the United States" because of environmental considerations. "One strain that is adapted to Chile, for instance, is relatively low in activity at temperatures below sixty-six degrees (Fahrenheit)."

Burris added that persuading such bacteria to thrive in a new environment will be difficult, "but it's rather simple compared to putting nitrogen nodules on corn. To convert corn, not only do you need bacteria, you also need to put in a type of hemoglobin which is now found only in legumes. You have to have the hemoglobin to supply oxygen to the bacteria, enough to make the ATP, but not so much that it will inactivate the system."

It may be possible to rearrange those free-living bacteria genetically to they become able to work at lower temperatures. Scientists working to decipher the riddles of cancer have already made, for example, viruses that have sensitive temperature "switches" that "click" into action at precise temperatures. So it may be possible someday to alter the temperature preferences of some of these nitrogen-fixing organisms.

Burris and his fellow scientists also noted, during an interview at Asilomar, that Dobereiner's Brazilian bacteria were finally recognized for their nitrogen-fixing abilities only in the past few years, even though they were first discovered in 1925, rediscovered in 1950, and then re-rediscovered in 1974. Eventually specialists recognized the potential of these organisms and got excited about them. So far, Burris said, Dobereiner has isolated about 100 different strains of these bacteria. "I think," he added, "that in tropical [farming] systems, at least, it is bound to be useful, and of course it has possible applications in temperate zones too."

This would suggest that efforts to transplant these creatures to the north may be one approach that is far ahead of the more difficult possibilities for nitrogen fixation. Still, some of the other approaches may soon make extensive use of the tricks of genetic engineering and move along rapidly, too.

Another researcher—Dr. Harold Evans at Oregon State University—has come across some hints that a few particularly vigorous plants, such as certain types of wheat and oats, may in fact already have some ability to fix their own nitrogen, probably through association with one or more types of free-living bacteria. If true, and if this trait can be manipulated, it could have enormous impact on agriculture around the world.

There are, of course, other approaches to the problems of fertilizer, and suggestions are coming from researchers indicating that gene transfers among bacteria—which are already being done —may provide the answer. Dr. Frank C. Cannon and Dr. John Postgate—both working at England's Agricultural Research Council nitrogen-fixation unit at the University of Sussex—report they've succeeded in moving the nitrogen-fixation gene, called the "nif gene," out of an anaerobic—or oxygen-hating—bacterium called *Klebsiella pneumoniae*, where it occurs naturally, into another air-tolerant bug known as *Azotobacter vinelandii*. Cannon and Postgate first transferred the nif gene into another bacterium, that favorite tool of genetics research known as *Escherichia coli* (*E. coli*), so that a plasmid—a small ring of DNA—could be made. Once the plasmid carrying the nitrogen-fixation gene had been constructed, it was extracted from *E. coli* and inserted into *Azotobacter*. As it turned out, the new genetic information carried by the plasmid enabled *Azotobacter*'s interior machinery to read and translate the nif gene, then carry out those genetic instructions for building ammonia.

This certainly indicates that the nif gene is transferable, and it also suggests that someday the nif gene will find itself living in the strangest surroundings, perhaps the least strange of which will be the world's most important food crops and the bacteria living with them. Such an accomplishment would be the most important step toward another Green Revolution.

The use of plasmids and bacteria, of course, isn't the only way to get those genes swapped around. Equally promising and equally difficult to achieve is the technique known as cell fusion. The trick is to take single cells from two different species, place them together in the same chemical bath and see if, when they touch, they can be made to join, to make a single new cell.

Already, in some instances, cells as different, as distantly related as mouse and man, as tobacco plant and man, have been induced to fuse and form new cells which were neither man nor mouse, neither plant nor animal. Indeed, those plant-animal hybrids are already being called "plantimals." Such strange new hybrid cells haven't yet been grown up successfully into strange new creatures, but similar experiments involving plant cells only have truly gone that far. Dr. Peter Carlson and his colleagues at the Brookhaven National Laboratory, at Upton, Long Island, New York, have been able to fuse the cells from two varieties of tobacco plants to produce a true intermediate hybrid of the two parent plants. It was also at Brookhaven—and the same group of researchers—where the man-tobacco fusion was done. Still, a much more ambitious—and as yet unsuccessful—goal is the production of hybrids from much less dissimilar species. The goal is production of truly unimaginable new types of fruits and vegetables. This approach might also serve as one more avenue for moving that nitrogen fixation gene into other species.

Until recently, however, it has been virtually impossible to do anything like true cell fusion with plant cells. The barrier to this type of genetic manipulation was the difficult, tough structure called the cell wall which surrounds each plant cell. These rigid walls are one of the important characteristics distinguishing plant cells from animal cells. The cells taken from animals are essentially naked, since they are enclosed only by a pliable outer barrier known as the cell membrane. Plant cells, too, have cell membranes, which are ultra-important for both types of cells, but only the plants have that semi-rigid cell wall barrier that is made up of cellulose, lignin, and sugars.

As botanists and other specialists learned long ago, the cell wall in plants is mainly what barred the earliest attempts to work with individual plant cells. During the process of removing cell walls for experiments, the cells inside—the protoplasts—were usually damaged and could not survive.

In the early 1960s, however, Dr. Edward Cocking and his colleagues in Nottingham, England, discovered that plant cell walls can be simply and gently dissolved or digested away by special enzymes which are poured into carefully prepared solutions with the cells.

After much experimentation, the result was development of a reliable way to produce "naked protoplasts" which can be used in cell fusion studies similar to the way animal cells are being fused. Naked protoplasts also turned out to be viable, functioning cells, and now Cocking's method has become established as the preferred denuding technique in botanical laboratories.

What, then, might be done with naked protoplasts?

First—according to Dr. Fuad Safwat, at the University of Massachusetts, Boston, who has worked with Cocking in England —came efforts to test the potential, the growth potential of these naked single cells. Experimenters soon found that it's possible to induce naked protoplasts to regrow their cell walls and begin dividing to produce clumps of identical cells. They can also be made to keep on dividing until they produce a callus, which is a rough growth or lump of tissue. Each callus, in turn, with proper manipulation of nutrients, hormones, and growth conditions, can be induced to grow up to become an entire new plant, an exact copy of the original plant from which the naked protoplast came.

A logical extension of this ability to get single cells to grow into adult plants, or parts of plants, is already being pursued. The group at the GTE Laboratories in Waltham, Massachusetts, is trying to trick cells from fruits and vegetables into maturing without going to the bother of growing adult plants. Through use of hormones, along with carefully controlled doses of nutrients, precise amounts of light and the right temperatures, the Waltham researchers have been able to grow ripe tomatoes without help— or with only very little help—from the tomato plant itself. It was explained that a tomato flower, grown on the young plant, can be snipped off even before it opens, before it can be pollinated. It is then placed in a nutrient solution, with the proper plant hormones, where it is coaxed to grow into a nice red, ripe tomato. An interesting sidelight is that this pretty fruit grows up with one quality that is strangely different: no seeds. This, of course, is what should be expected, since the original flower was pulled off the plant before it could be pollinated, and without that important dose of pollen no seeds can be formed.

Similar tricks are being played on the cells from soybeans. From experiments done so far it is evident that cells taken from the protein-rich edible tissue—the seed's endosperm—can be placed in a nutrient medium and be made to grow, producing more of the valuable protein found in soybeans. While this technique hasn't been developed on any sizable scale yet, there is the prospect that someday people will see huge factories pouring out soybean endosperm—for conversion into soybean meal—at the end of a long, slow-moving conveyor belt. In such a factory, the small snips of soybean tissue would be prepared and loaded onto the belt. As they moved toward the far end of the factory, they would be doused with nutrients, hormones, and light. By the time they reached the end of the trail they would have multiplied many times to produce large amounts of protein-rich soybean cells. No plants need be involved, except for supplying the first few bits of starting tissue.

Soybeans are already being widely used around the world as a source of protein, and in advanced countries such as the United States soybean tissue is being processed in amazing ways to serve as a substitute for meats and other staples. Thus, if soybean meal could be produced in even more massive quantities, cheaply enough and with scant use of energy, the results could make a big difference in the food picture everywhere. Since the whole procedure would be done indoors, such a system—a factory rolling out tons of soybean meal—would also be isolated from the vagaries of weather, from the problems of crop diseases, poor soil quality, and insect attack. Thus, by eliminating the need for the plant, it's also possible to eliminate the need for sunshine and soil.

But back to the topic of cell fusion. The second thing scientists hoped to learn through such experiments was something about the cell wall. They were interested in whether the naked protoplasts, so exposed, would try to build themselves new cell walls, and how quickly. They found that cell walls are soon regenerated. Also, while the protoplast was naked, it tended to assume a spherical shape that was relatively stable. They also found that the cells seemed to be in a frantic hurry to regrow their protective walls because scientists interested in fusion experiments had to move quickly to avoid new cell walls being put in place before experiments began.

The next step—taken in the mid-1960s—was to look closely at the possibility that two naked protoplasts might fuse when pushed together. Early experiments showed that some types of naked plant cells tended to fuse easily—even spontaneously—while other kinds of cells managed to hold themselves back from fusion.

Some tricks with fusion had already been performed by then with animal cells—such as joining the cells of mice and men—but the resulting hybrid cells haven't generally been viable for long, probably because of the genetic confusion created by the joining of two such dissimilar cells. In the chaotic aftermath of such a cell-fusion experiment, some of the chromosomes in the newly created cell appear to be gradually thrown out during subsequent cell division episodes, probably as the new cell begins sorting out which side—mouse or man—is going to be in control. As chromosomes are gradually thrown out, researchers can perform tests which tell them what properties are being discarded, giving hints about what genes are located on which chromosomes.

The scientists trying to do cell-fusion experiments, both with plant cells and animal cells, found eventually that some sort of helper is often necessary to induce fusion. In the animal cell experiments, for example, workers found that adding a disabled or crippled virus called the *Sendai* virus itself to the culture medium helped coax the cells to fuse. The virus itself apparently plays no role in the fused cell, but the discovery has helped establish some useful new lines for cell research.

In the plant cell experiments, however, *Sendai* virus was of no value, so fusion researchers began looking for other substances which could promote fusion. One chemical—sodium nitrate—was finally found to work in fusion experiments, making naked cells crowd together in a clump, and some of these, when squeezed together in a centrifuge, were eventually made to fuse. Later, however, a second chemical called polyethylene glycol (PEG), was found to be much more effective in promoting cell fusion, and it has become the more widely used of the two. PEG appears to be effective either with plant cells or with animal cells, and this discovery led to the fusion of these two kinds of cells.

The first viable, fully functioning new plants to emerge from these early cell fusion experiments were those hybrid tobacco plants mentioned earlier; the fusion experiments completed by Carlson at Brookhaven. He demonstrated convincingly that it is possible to fuse two naked protoplasts from slightly different plants and get new plants that are genetically midway between the two "parents." As also mentioned, Carlson's new plants turned out to be near-perfect intermediate hybrids in appearance as well as in genetic characteristics.

The plant-animal cell fusion experiments were done after Cocking, Safwat, and other members of the British research team decided to see if they could really trick the animal cells into mating with plant cells. What they used in this work were the red blood cells from a chicken—hen erythrocytes—and the naked protoplasts from yeast.

"My colleagues joked, suggesting we were really trying to make green chickens," Safwat recalled, "but we demonstrated for the first time the fusion of plant and animal cells."

Safwat explained that not all of the hen and yeast cells thrown together in the test tube chose to fuse. Some of the naked yeast protoplasts fused with each other (yeast to yeast), and some of the hen erythrocytes fused to their own kind (chicken to chicken), but still a small percentage of the chicken cells did join with yeast cells to form the strange new hybrid cells. Safwat added that no attempt was made to grow these fused cells then—in 1975—into some strange kind of creature, so nothing was known about what the results might be. Such plant-animal combinations—now being referred to as plantimals—would seem to be extremely difficult, if not impossible, to nurture into adulthood. Still, the possibilities inherent in cell-fusion experiments are intriguing, and perhaps a little frightening.

Safwat also noted that scientists are very interested in how the chromosome-filled nuclei of the joined cells sort themselves out after fusion; how they behave and make adjustments when they find themselves afloat in a strange new sea of cytoplasm. Do the nuclei themselves fuse, or do they both maintain their separate identities? One can imagine that both nuclei, each programmed and in the habit of being in charge, would find it difficult to share command inside a single cell. Indeed, it would tend to resemble two admirals on the bridge, each in charge of the same fleet, perhaps giving contradictory orders. One way to discover who's in charge is to watch them fight it out, monitoring how long one group of instructions takes to gain full command. In the plant-animal hybrids, too, one early clue would be the hybrid's beginning to make a new cell wall, which would hint that the plant nucleus might be gaining the upper hand, or at least that it still has some say about what gets done.

Of more immediate interest—especially in terms of results which might be applied to the world's food problems—however, are the cell-fusion efforts that remain wholly within the plant kingdom, new fusion experiments with types of food plants that are widely separated genetically. In Canada, agricultural scientists now doing pioneering work in cell fusion—at the Prairie Regional Research Laboratories in Saskatoon, Saskatchewan—have been trying to combine the cells of soybeans and corn, soybeans and peas, soybeans and tobacco, and carrots with barley. They haven't reported success yet in growing any adult plants from these odd crossings, but they have induced some of these hybrids to at least undergo cell division, produce tissues and even grow up to the callus stage.

One of the earliest goals of such work—as should be obvious with the emphasis on soybeans—is to try to make the nif gene settle into a type of plant where it's not normally found. The payoff from such a change could be dramatic, and it may not be as far off as many people suspect. Still another goal, however—one that will probably not arrive as soon as the transfer of the nif gene does—is to change what are called C-3 plants into C-4 plants.

The designation C-3 and C-4 doesn't really mean much to the layman, but to agricultural scientists it makes a world of difference. Plant scientists discovered years ago that the majority of the world's plants—including important crops like soybeans—are C-3 plants. This means that their photosynthetic processes—the physical and chemical pathways by which the plants use light from the sun for energy to build up carbon compounds—are relatively inefficient. The C-4 plants, on the other hand, include many of the fast-growing plants like corn and sugar cane, plus most of the other grasses. They are inherently more efficient than C-3 plants, by about 20 percent, because the C-4 plants are better at using up all of the carbon dioxide—and all of the energy—that they get.

According to Burris, from the University of Wisconsin, the C-3 plants use a metabolic system which takes in molecules of carbon dioxide from the air, combines each carbon atom with a five-carbon sugar molecule, then makes a six-carbon sugar molecule. This six-carbon sugar then splits to yield two molecules bearing three carbon atoms each. At this point, the C-3 plants rapidly oxidize—or burn up—these three-carbon molecules back to carbon dioxide. Unfortunately, they neglect to recover surplus energy, meaning that some of the energy temporarily locked into these molecules is released and wasted.

C-4 plants are more efficient because the plant cells begin work with a three-carbon acid, then add carbon dioxide to make it a four-carbon acid. The energy thus stored isn't expended too wastefully, so the C-4 plant doesn't need to work as hard, doesn't need as high a rate of photorespiration.

One goal of genetic engineering, then, will be to find some way to change the C-3 food plants into C-4 plants; in essence, boosting the C-3 plants a notch or two up the evolutionary ladder. (Plant research specialists believe that the C-4 plants represent a higher level of evolution than their C-3 cousins.)

The two basic approaches to plant science—adding the nif gene to nonleguminous plants, and the efforts to improve photosynthetic efficiency of the legumes—appear to be converging. One would suspect that Mother Nature, given time enough, might someday bring these two phenomena together anyway. Thus genetic engineers, when they finally get the job done, may just be beating Mother Nature at her own game.

But that's not all. A few scientists suggest there's an even better idea which may come to fruition via the cell-fusion experiments or through some other means of swapping plants' genes around. Botanists have frequently cast an envious eye onto a whole family of plants known as the succulents, which have rather thick, fleshy, juicy leaves and stems. The point is that these plants have developed a very efficient way of using whatever water they can get.

According to Dr. James Bonner, a biologist at the California Institute of Technology, "The succulents have a different basic strategy of operation, different from all other plants. Most plants, when they wake up in the morning and see the sun, open their stomata. These are the little holes, or pores, in the leaf surface where carbon dioxide can get in. But they also let the water leak out. And when it gets dark again, the plants close their stomata.

"In the succulents, however, the stomata are closed in the light and open in the dark. So what the succulents do is fix carbon all night. They gobble it up, using the C-4 enzyme system to make the carbon dioxide into organic acids. During the day they close their stomata and, with the energy they get from the sun, they break down these C-4 acids and fix carbon dioxide into carbohydrates using the C-3 pathway. Thus they use both types of enzyme pathways, C-3 and C-4, but at different times.

"Now it's true that the succulents don't take in as much carbon dioxide per unit of time as other plants, but on the other hand you can get ten times as much dry weight of plant per unit of water used. Right now, the only kind of succulent that is recognized as a crop plant is the pineapple. And that's why, in Hawaii, you see the irrigated crop—the sugar cane—in the lowlands. In the unirrigated uplands it's all pineapple because pineapple can take drought. So the obvious thing to do is to transplant the whole succulent type of metabolism to cereal plants so you can grow cereals in places with rainfall, say, of only five inches per year."

If suggestions like this can be taken seriously—and some biologists have been taking them seriously for years—it's obvious that all of these tricks with genetic tinkering mean we may be in for a truly startling revolution in agriculture. Indeed, if even a few of these ideas are realized in the field, where the food is grown, there will be significant improvements in production and the whole world food picture may change.

It means basically that we can actually expect some miracles, and they'll be coming straight from the laboratories where scientists are already tinkering with the chemical blueprints of the world's important food crops.

5

Meat: More for Man

FANS OF the Sunday funnies will remember that cute little creature, the Shmoo, which emerged—courtesy of cartoonist Al Capp—from Li'l Abner's mythical Valley of the Shmoon. They'll recall, too, that the shmoo was a satirically comical, whimsical answer to the world's growing, unfunny food problems. Shmoos yielded steaks, chops, and roasts on demand; happy to do so. Shmoo whiskers—the only leftovers—served as ideal toothpicks for the folks who'd just polished off a shmoo or two.

Al Capp's ideas were worth some laughs, but perhaps in an age of genetic engineering they won't be all that farfetched.

The cartoonist probably overstated the theme a little, but before long some complex and versatile food animals can be expected to come from the laboratory work of scientists who are already playing with animal genes and live animal cells. Someday they're going to create new and better types of food animals to help fight the world's increasingly severe hunger problem. Meat animals, of course, along with fish, poultry, eggs, and some plants, provide some of the world's most important sources of protein.

Right now, unfortunately, the only creatures coming close to mimicking the shmoo's versatility are the single-cell organisms which can be grown rapidly and then be transformed into interesting food products in ways similar to what's now being done with soybeans. Better yet, however, these single-cell protein organisms can probably be rather easily manipulated genetically to more closely fit the needs of man.

It won't happen immediately, but it's not difficult to imagine scientists beginning soon to insert new and different genes into single-cell organisms—such as algae—to add or change color, to alter the odor, the flavor, consistency, or even the vitamin content of the finished product. The genes that can code for such qualities already exist in one organism or another, and it seems reasonable that one day special genes will be taken from their natural hosts and will be passed around for use in other creatures. Far in the future, indeed, it can be expected that plant and animal genes might be isolated and passed back and forth to create whole new classes and types of organisms, whole new species that are able to grow, reproduce, and be harvested even in areas where severe environments make food production nearly impossible. Already plant and animal cells are being fused; scientists refer to these strikingly unnatural hybrids as "plantimals."

Researchers will bump up hard against some limitations to such techniques. Even if experiments show that new types of organisms can be designed and grown, in some cases it won't be economical to raise them, or there may never be a market for them. Indeed, many potentially important meat animals exist already, but because of traditions, local taboos, economic problems, or other reasons, they are woefully underutilized or not even harvested at all. A good example of this is the squid, a somewhat repulsive creature found in abundance in American waters but largely neglected except for use as bait by fishermen. By comparison, in Europe, the Middle East, and Asia, the squid is a prized, valuable food which supplies good amounts of protein to large populations.

American diners, however, have mostly avoided this protein-rich, delectable food even though it is readily and economically available. Some work is being done to overcome this problem, notably at the Massachusetts Institute of Technology, where under a U.S. Sea Grant program efforts are being made to find some way to bring the slimy little squid into favor, to appeal to fussy American palates. One suggestion—which seems to have possibilities—is to slice the squid meat into small rings, similar in size to onion rings, and sell them through existing fast-food establishments like McDonald's or Burger King. This could probably be done easily right now except that federal regulations prohibit selling such foods under assumed names. So it will still have to be called squid, and the name's enough to turn some people's stomachs. It's an unfortunate fact, too, that nobody is going to be able to change a nation's tastes quickly; it's not something that's accomplished overnight.

Similar problems of acceptability can be expected when new foods that have arrived through genetic engineering finally hit the market. Thus it can also be expected that the new foods will be developed and marketed so they closely resemble foods that are already popular. Right now we're seeing food products—remember Hamburger Helper—that mimic or blend with existing foods, or which "extend" accepted foods to make them serve more people. Soybeans, too, are being treated—by spinning, for example—to provide some close cousins for some foods already in use, especially meats.

What's predictable, then, is that genetic engineering will begin by adding important new substances to the existing market for semiartificial foods. But it can also be expected to lead to the improvement of future meat animals and to probably bring other types of animals into the realm of food production. It's not difficult to imagine some of the possibilities, some of them more believable than others, such as:

• New, different types of milk, either low in fat or high in fat but low in triglycerides, or flavored in new ways. Vitamin D, too, might be built in genetically rather than being added later during processing, as is now the case.

• New types of meat, probably low in fats, that are enhanced in flavor, texture, and are less susceptible to spoilage.

• Fish and fowl with new flavors, colors, and aroma, which will probably also be more nourishing. Already, at the University of California at Davis, and at the University of Maryland, scientists have discovered and are analyzing a mutant strain of chickens which never grow any feathers.

The main appeal of such naked birds is that more of their food energy goes into building meat rather than feathers. Researchers, however, have also found that a featherless chicken is a cold chicken. The naked birds, unless given special care, die quickly because of heat loss. On the other hand, the bald birds tend to survive heat waves in which other chickens die. At present the main drawback is that featherless chickens must be kept indoors in chicken houses heated close to 80° Fahrenheit, which is expensive. In addition, these featherless birds burn up more of their food energy just trying to keep warm, perhaps wasting as much— or more—food energy than was saved by not growing feathers. Calculations done by the researchers, however, indicate that about 25 percent of a normal chicken's live weight is feathers, which are rich in protein. The scientists found that while a chick is developing, the growth of feathers has a high priority, with feathers gathering up the essential amino acids which make proteins, even taking precedence over the whole bird's rate of growth. Thus a naturally naked bird tends to make more efficient use of the amino acids in its food during this early growth period than do birds which pour the energy into building feathers. The difference is seen in terms of greater weight gain during early weeks in the chicken's life.

There are, of course, other problems. Taste tests indicate these naked birds are as savory as their more normal brethren, and they tend to lose less weight in cooking because their muscle tissue contains less fat. The main problem, however, is that aside from having to keep such birds extra warm, they come into the market with smooth skin, which is expected to be rather off-putting for shoppers who come along looking for the normal chicken's de-feathered, dimpled skin.

Naked chickens, however, are more representative of the traditional crossbreeding techniques now in wide use, not yet approaching what is considered true genetic engineering. So where are we, then, where do we stand, in relation to true manipulation of the genes of food animals?

Basically, no completely new types of food animals appear to be under development now, and no genetic engineering is under way yet that will soon produce any important, dramatic changes in present crop animals. Breeding and crossbreeding are still—and will continue to be—most important, and for the next few decades will obviously be the main source of change in modern livestock.

Nonetheless, important work is well under way now in which veterinarians are coming ever closer to the ultimate goal of being able to clone up whole herds of certain desirable types of cattle, sheep, or other food animals. The basic idea is to build up herds of ideal animals in which all are genetically identical. The techniques being developed by veterinarians, however, are still closely associated with the practice of artificial insemination, so the newborn animals are the product of egg and sperm joining, and reshuffling of the genes. As it's being done now, however, the time doesn't seem too far off when sperm and ova will be joined in the test tube rather than in the female's womb, producing living embryos which can later be implanted into a "host mother's" uterus to grow to full term. Some veterinarians, too, are already using one interesting technique to "amplify" any particular set of genes that has been produced naturally when a bull's sperm has been inserted into a cow's reproductive tract.

As explained by a group of veterinarians now practicing in Montclair, California, American dairymen in recent years became greatly interested in what are considered exotic breeds of cows from Switzerland, Austria, Italy, and some of the African nations. Unfortunately, because hoof-and-mouth disease is prevalent in some overseas areas, U.S. regulations prohibit the live importation of these animals. But Canada has set up some quarantine stations for use in importation of livestock, so these cows can at least be brought into North America. Bull semen from abroad can be brought freely into the United States.

Thus eventually a few of these full-blooded exotic cows were brought into the United States via Canada, but not nearly enough of them arrived to satisfy the demand. It then became economical to use a relatively new technique, an amplification technique known as ovum transplant.

This work involves dosing a desirable cow with an injection of a horse hormone—pregnant mare gonadotropin—which causes several follicles, or ovum (egg) sites, to form or ripen on the cow's ovaries. The cow then receives a dose of a human hormone—human chorionic gonadotropin—which causes ovulation, or break-out of the ripe ova, and formation of the corpus luteum, which is the hormone-producing cavity left in the ovaries when each egg is released. Once ovulation has occurred, the veterinarians arti-ficially inseminate the cow, injecting 10 million to 15 million sperm cells from a preselected, full-blooded prize bull into the cow's uterus. These sperm mate with the eggs and begin to form embryos, which will grow up to become adult cattle.

Five days after insemination the cow is anesthetized, an incision is made in her abdomen and her uterus is "exteriorized," or pulled out where it can more easily be worked on. Then the fertilized eggs are "harvested" by one of several techniques such as flushing the uterus with a saline solution. Once the developing eggs have been washed out and collected, they're inspected under the micro-scope to carefully select those which have been fertilized. Those that have accepted a sperm cell are sorted out individually and are inserted into the womb of another anesthetized recipient "mother" cow. From then on, each embryo, well-nourished, warm, and comfortable inside a healthy cow's uterus, is carried like a normal pregnancy all the way to term. At birth the result is usu-ally a perfectly normal calf which is unrelated to its host "mother." What it is is a careful product of a prearranged mating of a bull and cow which never met in the field, and still another cow which was made to suffer through the pregnancy.

This won't sound like much fun for the bull—whose sperm was probably shipped, frozen, from some distant farm—or for the donor cow, or for the recipient cow, but it makes sense eco-nomically in some cases, especially with rare, expensive breeds.

Dairy farmers are learning that through artificial insemination, use of sperm from prize bulls can improve the quality of their herds by almost 3 percent per year. And, by the use of this ovum transplant technique, dairy herds can be improved to even greater degree, faster.

The main drawback so far is that a dairy farmer must want one of these exotic animals very badly; the cost of performing these ovum transplants is very high. On average, one group of veterinarians doing this work at Colorado State University, Fort Collins, was selling each pregnant cow carrying an exotic embryo in her womb for about $1,500. This may seem high for an ordinary cow carrying an extraordinary calf, but the California veterinarians pointed out what causes most of the expense. The ovum transplant technique requires that the recipient cows—those ersatz mothers—be in exactly the same stage of the twenty-one-day estrus cycle as the donor cow. Indeed, there must be at least ten or twelve recipient cows ready in the same stage of the estrus cycle to receive all the eggs produced by "superovulation." Thus if a veterinarian intends to transplant ten fertilized eggs from one prize cow into recipient cows, he must maintain a herd of 210 cows at all times. Obviously, maintaining such a herd is expensive, and the group at Fort Collins, as an example, kept a herd of about 600 recipient cows ready for ovum transplants.

The California veterinarians noted that they had earlier been involved, individually, in ovum transplant work, but that American dairy farmers' interest in the exotic foreign cows has waned and the ovum transplantation business has been largely closed down. At its height, though, especially as practiced at Fort Collins, the transplant system was considered successful and productive. In a month's time, under ideal conditions, one prize cow could produce, with the aid of recipient "mother" cows, as many full-blooded calves as she could normally produce in her entire lifetime. Apparently the only limiting factor in the whole process was the surgery done on the donor, rather than the drugs involved, because after numerous operations to "exteriorize" the uterus, the cows tend to develop adhesions.

One smaller problem was encountered, too, which was that in some transplant attempts, for some reason, not all of the fertilized eggs were flushed from the donor's uterus. In such cases the donor herself becomes pregnant, which puts her out of the donor business for months until her calf is born.

Specialists have commented, too, that even though the ovum transplant business has mostly disappeared, it can still be an important technique if used on a smaller scale. Domestic farm animals such as the cow never really reach what is known in humans as menopause, so even after a cow's major reproductive organs are no longer capable of carrying offspring, her eggs are still produced as usual and can be harvested and distributed to other cows.

Actually, this smaller-scale approach may be of more value in the long run than the efforts to reproduce a lot of exotic cows in a hurry. This is true because only after a cow has fully matured— after she has produced several calves and has gone through several lactations—are farmers able to assess accurately what kind of milk producer she will be. In the rapid production of exotic cows the work was done with heifers, rather than fully mature cows, so the dairymen who bought the calves really knew very little about what the grown cow would do for their herds. In the future, it's most probable that ovum transplantation will be done more often as a means of maximizing milk production, mainly by selecting for those genetic qualities that lead toward more or richer milk.

So far, these ovum amplification techniques have been mostly limited to cows. Horses, obviously, might be better candidates for such work since they are frequently expensive enough to warrant the costs involved. Unfortunately, however, veterinarians haven't yet discovered the hormones needed to make a mare superovulate. There is still a chance that some man-made hormones will be able to do this. If so, the ovum transplant business should begin to flourish. Given the prices paid for thoroughbred racehorses, quarter horses, and specialty breeds, there's little doubt that ovum transplantation would quickly find a rich and ready market.

An important finding already coming from studies of horses, however, is that when it comes to breeding to produce new colts, veterinarians seem to do a better job than the stallion can; artificial insemination is a considerably more successful method of colt production than natural mating. Veterinarians have discovered that in some valuable mares the male's sperm—injected in massive amounts by a stallion—can cause an immune reaction in the uterus so that implantation of the embryo does not occur and no colt is born. By injecting smaller, better-controlled amounts of sperm, veterinarians can avoid this defensive reaction and thus get an otherwise barren mare to bear a colt. Some mares, unable to conceive naturally, have been enabled to produce colts solely because of artificial insemination.

The veterinarians expect, too, that once the proper hormones are available so they can do ovum amplification in horses, the process will be even easier to perform in mares than in cows, since a mare's cervix is easier to dilate than a cow's for implantation of the growing embryo. They mentioned that this is also true of humans, and said there is some suspicion that transplantation of embryos has already been done, at least experimentally, in women, but without publicity.

The science involved in the transplantation of living embryos has, of course, moved rather far ahead of these more commercial applications. It has already proven feasible, for example, to take a fertilized cow's ovum—the embryo again—and place it inside a rabbit's womb and use the rabbit as a convenient, living transport vessel. So far it has been found that a living bovine embryo can survive for more than eight days inside a rabbit's uterus, a period that is long enough for the rabbit to be transported by air across the Atlantic Ocean. Through this approach, living embryos tucked away safely inside rabbits have been flown from the United States to Scandinavian countries—and from Austria to Oklahoma—to be implanted into recipient "mother" cows. At present there is little doubt that such a transportation system, if developed thoroughly, could be used routinely.

Hiding a bit of bovine treasure inside a bunny rabbit, however, isn't the only method scientists and veterinarians have developed for transporting live embryos. Another approach is through cryogenic freezing, cooling the undeveloped living creature down to the temperature of liquid nitrogen ($-196°$ Celsius). This allows the embryos to be stored successfully for much longer periods. It's difficult to imagine how a complex, living, fully viable organism is able to survive such potentially deadly treatment, but experiments have shown that it can be done successfully. A research group at the Jackson Laboratories, in Bar Harbor, Maine, has already shown conclusively that living mouse embryos can be frozen down to $-196°$ Celsius, can be stored, and can then be thawed and placed into a receptive mother mouse's womb to grow toward normal birth. In addition, these frozen mouse embryos have been shipped successfully from Bar Harbor to England, where they were thawed out, implanted in mother mice, and then grown to full-term birth. Work in England is showing that these deep-freeze techniques are also possible with embryos from the most popular meat and dairy animals.

These small step-by-step developments seem to be leading to the time when veterinarians will be able to scoop the nucleus from one cow's ordinary cells, such as a skin cell, transplant it into an unfertilized egg—thus turning the egg "on" with a full set of chromosomes—and begin growing up a new identical copy of the original cow. Too, before this embryo gets very large, perhaps at the sixteen-cell blastocyst stage—these cells can be dissociated, broken apart from each other, and each can then be put into a recipient cow to grow to term. This, of course, is "cloning," and it will eventually provide one way for perpetuating especially valuable gene combinations.

The next step beyond cloning, certainly, will be to actually start manipulating individual genes—as is already being done with the genes of bacteria and viruses—to purposely build up animals with valuable gene combinations that can be grown up in multiple copies through cloning. The goal would be to produce whole herds of perfectly matched cows, pigs, horses, or other farm animals, all blessed with identical sets of genes. Of course, it's also true that this technique may also be used with human tissues, but for now there doesn't seem to be much advantage—or profit—from producing people this way.

One important problem facing the future of animal science, however, is going to be the availability—or lack of availability—of those special desirable genes coding for things like disease resistance, use of food, and stamina. Right now among the world's domestic and wild animals an enormous gene pool exists, but specialists are worried that some of the older, purer, and more obscure animal breeds will disappear as the more productive American and European hybrids are introduced into herds in the developing nations. This problem is almost identical to that facing plant breeders who are seeing the primitive types of food crops like wheat and barley displaced—even made extinct—by the encroachment of more commercially desirable hybrids.

Unlike the substantial efforts now going into the preservation of plant species, however, only small fledgling efforts to preserve important animal stocks are now under way. Fortunately the United Nations Food and Agriculture Organization (FAO) is starting to work on this problem, and many, many years from now FAO may finally have a viable program going.

If so, then some of these obscure types of food animals—and especially the cattle—may still be available for conventional cross-breeding experiments, and their genes will also be available for advanced genetic recombination studies once the craft of genetic engineering is a reality. Between these two extremes, certainly, it can be expected there will be a period when those tiny embryos will be taken from the cow's womb—or even be produced in test tubes—and then be frozen for long-term storage. This, too, will become an important method in preservation of the worldwide gene pool.

In the meantime, however, the only alternative is to set up small, controlled breeding herds of the most interesting animals, especially those in the developing countries. These animals are still relatively unstudied, and they may carry important genes for better meat quality, higher milk production, and disease resistance.

One such program is already being started to help preserve a strange breed of humpless African cattle called kuri, inhabitants of the islands and shores of Lake Chad. The most obvious unique characteristic of the kuri is that they bear an odd set of inflated, spongy horns. No one has yet figured what evolutionary purpose these horns serve, but it is probably related somehow to the fact that kuri are excellent swimmers. They normally forage in the lake, consuming the coarse water plants growing in shallow areas. In addition to this strange set of horns, the kuri are also known for the excellent quality of their beef. They produce only moderate amounts of milk, however.

The problem facing the kuri as a distinctive breed is that their gene pool is being diluted, or even lost, as they are gradually crossbred rather indiscriminately with other types of African cattle. At the same time, their habitat along the shores of Lake Chad is being destroyed as the water level rises.

Because of this fast rate of change, in 1972 the Lake Chad Basin Commission proposed that a new project be started to conserve what was left of the kuri. This proposal was meant to serve several purposes. First was to preserve a type of cattle that is different enough to be a tourist attraction. Second was to demonstrate that conservation of obscure breeds is possible even in the less developed countries.

It's unfortunate, however, that the kuri are only one out of hundreds of breeds that deserve preservation for future study, and it is possible that some of those neglected will soon drift into extinction.

Ideally, programs like the one being run at Lake Chad should be duplicated in Asia, in the Middle East, South America, Europe, and even in America. Such efforts should be expanded much beyond the preservation of cattle only; encompassing other domestic animals like pigs, sheep, goats, chickens, ducks, turkeys, geese, and even fish.

Again, however, in contrast to the expanding systems for the preservation of plant materials, these animal preservation programs tend to be vastly more expensive. Thus, until the technology for long-term freeze storage is developed to a commercial scale, real progress toward saving many rare animal breeds can't be expected.

Nonetheless, it should be evident that expanding efforts to enhance the growth, protection, and production of food animals is becoming one of the most important areas of genetic research. And it shows no signs of slowing down.

We can expect that enormous strides toward solving earth's chronic protein shortage will be made through improvements in existing animal species. And in the long run it seems logical to assume that some wholly new types of animals will emerge from experiments, most of them probably coming from the laboratories run by genetic engineers.

Indeed, if scientists ever get around to creating a few real samples of Al Capp's lovable shmoos, the chances are we'll be in clover—except that people's diets might become more rich than healthy.

6

Tomorrow: Living Factories

WORD ARRIVES at 4 A.M., jolting a groggy seaman out of his late-watch stupor. The clattering teletype prints out the following message:

> URGENT. Commercial tug, *Janice*, 6 oil barges in tow, reports cable parted, 2 barges aground on Catalina Island, 4 adrift. Each carries 600,000 gallons light crude. One barge ruptured. Oil believed leaking since midnight. Expect 14-mile slick by dawn. Please advise.

Massive pollution is loose at sea, spreading slowly, silently, mile upon mile downwind. Oil, the lifeblood of industry, has spilled again, befouling the sea, wreaking havoc on the fragile ecology of the deep.

Unfortunately, not much can be done now other than scramble the U.S. Coast Guard's emergency oil-spill units to try, first, to stop the leak; second, to slow the spread of oil; and, third, somehow sop it up.

But tomorrow—and it's not a very distant tomorrow, either—this may be a very different story. Oil will still be spilled and it will still flow inexorably downwind or downstream, mucking up everything it touches, but the response will be different. The oil will be gobbled up before it travels very far.

Who's going to eat it?

Bugs—tiny microbes designed and built in a laboratory—especially tailored with a taste for petroleum and hungry for a meal of oil. And it's not science fiction.

The weapon? It's a new version of an old bacterium called *Pseudomonas*. Now it has a built-in appetite for oil, and it represents the first obvious commercial application of the embryonic craft of genetic engineering. The bug has now been built and just waits to be put to work.

This task was done—and the work patented—by Dr. Ananda M. Chakrabarty, a research scientist at the General Electric Company's research and development center in Schenectady, New York. Chakrabarty envisions someday seeing his oil-hungry creatures sprayed as a powder directly onto a spreading oil spill. At present, he says, this new bug can gobble up about two-thirds of the hydrocarbon products found in crude oil, and a second eager oil-eating creature is being designed to handle the rest.

Better yet, the leftover products—the wastes—produced by *Pseudomonas* will serve directly as food for marine organisms.

Industry, worldwide, is already taking a strong interest in the continuing successes of genetic engineering; industrial laboratories are now looking carefully at some of the promising work and the new techniques coming from the academic laboratories. Some useful results from genetic engineering are closer to reality than many people suspect. Most of the interest for now involves microbes, and microbes are the easiest creatures of all to manipulate genetically.

Microbes—the world's largest invisible population—are already widely used by industry, so new techniques and new materials that can improve what is already going on will probably be quickly adopted. Indeed, the list of products and processes which depend on bacteria, molds, or their specialized enzymes is formidable. Some of the simpler ones include:

• Wine. Without living yeast cells to convert the sugar of the grapes into alcohol, this industry couldn't exist. A great number of very specialized yeasts have been found and are widely used, but the tricks of genetic engineering promise someday to provide yeasts that are able to work harder, faster, and more efficiently,

(G. E. RESEARCH AND DEVELOPMENT CENTER)

Redesigned bacteria. A tribe of microbes—thought to be the world's first commercially genetically-engineered bacteria—is tested in a puddle of oil, floating on water, at the General Electric Company Research and Development Center in Schenectady, New York. These microbes were specifically designed with components from four different strains of bacteria so they can efficiently gobble up the oil spilled on rivers, lakes, and at sea. This new oil-hungry "bug" was devised by General Electric's Dr. Ananda M. Chakrabarty.

turning more grape sugar into alcohol faster. At present, yeast usually dies—or at least stops the sugar conversion process—when alcohol content gets much above 13 percent. This, too, might be changed, leading to some naturally stronger wines.

• Cheese. An enzyme called rennet which is added to milk is the key element which causes milk protein to coagulate in the cheesemaking process. Until recently, this enzyme was extracted from the stomachs of suckling calves. Now Japanese scientists have discovered an enzyme made by a mold which produces a powerful rennetlike response in milk.

• Flavors. The Japanese have long used seaweed and a dried fish—the bonito—as flavor-enhancing substances for bland foods. In recent years they've learned to use a type of mold to produce these flavoring agents from wastes of the woodpulp industry.

• Hormones. Pharmaceutical companies have long taken advantage of the process of fermentation—in huge vats under carefully controlled conditions—as a relatively cheap, productive method for producing steroid hormones like cortisone.

• Waste treatment. Through digestion of sewage and degradation of other products, microbes are a key element in final disposal of the world's refuse. Degradation of wastes can also, in some cases, yield useful energy through production of methane gas.

Such examples, simple as they are, will illustrate the large stake that industry has in the abilities of tiny living creatures to do important jobs. They should also illustrate why industry—and especially industry in Japan, where the use of microbes is probably most extensive—is vitally interested in the whole field of genetic tinkering.

Even without genetic tinkering, however, there is still enormous potential for production of new substances just because of the overwhelming diversity of microbial life. Indeed, scientists have so far identified more than 2,500 microbial species, and some specialists believe this number represents only about 10 percent of the different types of microbes living on earth. Microbes are found almost everywhere, even in the most extreme, hostile environments. They have been found afloat high in the atmosphere, deep in the seas, buried in the ices of Antarctica, in the steaming waters of geysers, and even in the saline waters of desert hot springs.

Rapid reproduction is a microbial specialty, since some of these invisible organisms, given a healthy environment, can run through seventy generations in a twenty-four-hour day, and a small sample of nutrient medium can carry some 700 million individuals.

The interesting thing is that despite such promiscuity—and the existence of such vast numbers of individual microbes—during the greatest part of the history of the human race mankind had no idea such tiny living things even existed. Finally, in the latter half of the seventeenth century, a Dutch merchant and amateur scientist, Anton van Leeuwenhoek, using his own hand-ground lenses and a homemade microscope, first saw that a whole new world of unexpected organisms exists. His name for these strange "cavorting beasties" was "animalcules."

It took two more centuries, however, before Louis Pasteur finally demonstrated the important role these peculiar little creatures play in natural processes. Pasteur astonished his own world by proving that microorganisms are responsible for a vast array of formerly unexplained phenomena ranging from the fermentation process that produces beer and wine to a host of diseases which afflict man and beast.

Studies since then have added immeasurably to the store of knowledge called microbiology, but perhaps the most telling piece of information to gradually emerge involves diversity, the great number of very different organisms which share much of the same internal chemistry. Microbial species differ from each other in things like nutritional needs, in the foods they can break down and utilize, in their energy sources, and in the substances—the end products—they manufacture. The various different types of microbes are also known to depend on thousands of their own different enzymes, which are the potent biological catalysts which help important chemical reactions proceed.

Such enzymes are truly specific, each doing its own particular task, each promoting only one very special biochemical reaction. And, by growing, collecting, crushing, and separating a culture of living bacteria, scientists have been able to isolate specific enzymes, study them, and also reproduce—in the test tube—the chemical reactions these enzymes catalyze for the living microbe. Thus it has been possible, step by step, to begin unraveling many of the mysteries of the various metabolic pathways, the processes by which living things of all sizes break down and dismantle foods chemically for later use as the organisms' own building blocks, their raw materials.

Such work finally led to the important "one-gene, one-enzyme" theory which holds that each enzyme's structure—and thus its function—is controlled by the one gene which codes for it. And this is what is important for industry, especially as genetic engineering gets under way. Scientists are finding they can begin to control the microbe's interior machinery, that there are interesting mechanisms which allow research workers to manipulate what a bacterium does with its food and its energy.

The key, of course, was discovery of the plasmids, and the finding that these ring-shaped molecules can be snipped apart and recombined to make new genetic combinations. Perhaps more important, these plasmids—which are simply small rings of DNA (deoxyribonucleic acid)—are not an integral part of the bacterium's normal collection of genes. Thus the plasmids can be passed back and forth among bacteria like so much coinage, and a bacterium receiving a plasmid may be given the ability to outsmart a new antibiotic drug, or to produce some useful new enzyme. This suggests, then, that plasmids are merely a natural type of carrier for bits of genetic information which are added—when needed— to the bacterium's normal genetic bank account. Plasmids, however, are apparently *not* mixed in with the creature's normal set of genes on the chromosome, but float free inside the tiny cell, ready for easy exchange with other bacteria. And for industry, at least, this is what makes much of the difference.

Now that plasmids can be easily cut up and recombined, made to carry almost any genes the scientist wants them to, plasmids can be used as the vehicles to carry different genes into a wide variety of bacterial hosts, and the hosts in turn, as they divide, keep recopying those donated genes over and over again indefinitely. Thus the bacterium produces extra copies of the donated gene—which is in itself important to scientists—while building up a large colony of bacteria, all designed to do a special task. In genetic engineering for industrial purposes, the goal is to give some well-known, easily controlled organisms new pieces of DNA which induce them to perform jobs that can now only be done— or can't be done at all—by expensive, complex chemical plants.

This, then, basically, is the approach used by Chakrabarty for manufacturing his new oil-gobbling bacteria. Indeed, his genetically engineered hydrocarbon-hungry microbe now carries pieces of information donated on plasmids from four different microbes, each of which had its own natural, but lesser, ability to consume petroleum hydrocarbons.

The trick was to stuff these four different plasmids into *Pseudomonas* bacterium, which Chakrabarty was apparently able to perform through a bacterial mating process known as sexual conjugation. During this process two bacteria, side by side, exchange some of their interior chemicals through a tubular connection. Plasmids coding for specific reactions can flow through the tube from one bacterium to another.

Chakrabarty somehow managed to get four of these plasmids into *Pseudomonas,* but his employer, General Electric, declines to discuss it, saying his method is "proprietary information," a trade secret.

Anyway, Chakrabarty's new plasmid-rich bacterium was found to be somewhat unstable. It wasn't able to divide properly to pass on copies of all four plasmids to its offspring. Without reliable reproductive ability the new bacterium would be useless for oil eating, so Chakrabarty reports he overcame the problem by exposing his bacterium to powerful ultraviolet light, causing small changes in the plasmids that made the system more stable. The result was that the new bug's oil-eating talents are now routinely passed on to all its offspring.

As a result, Chakrabarty and General Electric now have a new custom-built bacterium that can eat up oil several times faster than other natural organisms can. When it will be used, however, is open to question, since field tests haven't been run and such tests may well prove difficult to perform. The main obstacle, probably, will be that the U.S. National Institutes of Health (NIH) have issued guidelines covering such genetic recombination research, and the rules prohibit the release of such newly created bacteria into the environment. Chakrabarty and General Electric aren't legally bound by the NIH rules, which apply only to recipients of NIH research funds, and also to funding supplied by the U.S. National Science Foundation. Still, if General Electric chose to ignore these rules and released the new bug into the world's water, this act would probably stir up such a fuss—even among Chakrabarty's fellow scientists—that it probably wouldn't be worth the fight. Too, General Electric's public relations program might never recover. Obviously, no one wants to be first to release a potential biological disaster on the world. Once the new oil-eater is turned loose, there may be no calling it back.

So, for the time being, at least, these new oil-hungry organisms have been put on the shelf to await further developments. In the meantime, work is continuing in several other areas. Chakrabarty also reports he's working on new bacteria which seem to have the ability to gobble up and concentrate precious metals like gold and platinum.

This, too, may prove important, since the world's resources of rich, easily mined and easily extracted metallic ores are being rapidly depleted. High-grade ores containing copper, nickel, chromium, tin, and molybdenum, are becoming increasingly scarce. At the same time, however, newer work is showing that certain microbes already exist which can be used—and are being used—to extract metals from ores once the ores have been dissolved in solutions.

This ability of microbes to grab and hold onto metals was perhaps first noticed in Wales and Spain, where water has long been used for leaching copper sulfide ores from the ground.

Some years ago, a species of bacteria, similar to a bacterium called *Thiobacillus ferroxidans*, was found in Spain's copper-rich Mians Riotinto region. This same organism was also found during studies in Pennsylvania coal mines, and it was also discovered in copper-laden water in the largest open-pit copper mine in the United States, Bingham Canyon, Utah. Subsequent laboratory research showed that sulfurous ores carrying eight different kinds of metal can be worked on by this bacterium.

As a result these bugs were put to work on the very low-quality ores left over from copper mining, the tailings which accumulate in huge piles near mining and smelting operations. And the process works. In 1966 the United States dumped 370 million tons of copper mine refuse which usually contains small amounts of copper metal. By then, however, microbial processing of these tailings was already producing some 10 percent of the nation's copper.

Microbial processing of ores—especially the low-grade ores—has become a reality. Economically the process seems to be viable, since it produced copper at a cost of about $1,000 per ton at a time when the world market price was $1,400 per ton. Thus the process of using living microbes to gather the copper from ores in solution is being expanded, with new work being done in Russia, Mexico, and the United States.

This process isn't limited to copper. *Thiobacilli* are also able to help with the extraction of uranium from low-grade ores, and it is expected this, too, will become an increasingly attractive method as the price of uranium for nuclear power plants keeps climbing. In Canada, for example, at Stanrock, some 7.5 tons of uranium have been extracted from the walls of underground mines. The microbes, already at work in the rock walls, are sprayed with the proper solutions and the resulting liquor, rich in uranium, is then pumped to the surface. Thus the miners escape the digging and carrying out of huge amounts of heavy rock in pursuit of small amounts of useful metal. The microbes do it for them.

In South Africa, too, mining specialists have been using bacteria to eat up or dissolve the gold found in what are called laterite deposits. These specialized bacteria need organic substances to be supplied as food, but they are still providing one important means for working low-grade deposits. In the Soviet Union during research at the Irkutsk Institute of Rare Metals, scientists found that within twenty hours, gold-eating bacteria can dissolve about 30 percent of the gold out of a given ore sample. South African researchers are saying they think some organic compound secreted by the bacterium is responsible for dissolving the gold out of the ore.

There will be several other advantages to using microbes as miners. First, they use up very little energy. Second, much smaller investments are required; and third, this provides a way to work back through the ores—the tailings—which were once considered too low-grade to waste time and money on.

The main drawback, so far, is that microbial ore processing is too slow. When the techniques of genetic engineering are finally applied, however, chances are this whole process can be speeded up, similar to the way the oil-eating process can be hurried along by General Electric's new microbe.

As a matter of fact, Chakrabarty expects that someday many of these scarce metals will be extracted directly from seawater by bacteria which have been specially prepared for the job. This, he said, will probably be true first for the noble metals like gold, platinum, and palladium.

"It's very difficult," he added, "to separate platinum from other noble metals. So our idea now is to isolate genetically engineered bacteria that can easily separate these precious metals." He added that he has already found a new bacterium that can consume gold, plus another bacterium that selectively sorts out platinum and palladium. Other types of bacteria have been found which can also withstand high concentrations of poisonous mercury.

Chakrabarty also noted that microbes' metal-handling ability appears to be carried—coded for, genetically—by genes on plasmids, so this means it may be transferable to other bacteria, and that extra plasmids can probably be shoved into a single bacterium to enhance its ability to process a metallic ore.

Scientists who've been looking at this strange metal-eating ability of microbes have come to suspect that it is an evolutionary response to the presence of poisonous materials, the toxic metals. In many cases, such as with mercury, metals in the environment tend to poison living organisms, and it seems that some bacteria—in the effort to carve themselves out a comfortable niche in the wild—developed ways to circumvent this problem. One way to do this—for the microbe, at least—would be to grab onto each metallic molecule that comes along, swallow it and cover it with a protein coat, then tuck it away in a corner where it can do no harm.

When asked how one gets the precious metal out of a bacterium once it has been put safely away, Chakrabarty answered that the process is simple. Just heat up a batch of microbes until everything but the metal has boiled away.

Metals aren't the only industrial products being considered as candidates for microbial processing. Right now some agricultural researchers are looking at food-producing bacteria as a means of breaking down some of the complex sugars in soybeans that the human intestine can't handle. The result, they hope, will be a new type of soy milk which researchers are confident can then be turned into a new type of yogurt. Yogurt is made by fermentation, calling for the use of another type of microbial organism. Flavored yogurt from this material is also possible, according to the U.S. Agricultural Research Service, and can be made by simply adding vanilla, orange, strawberry, or lemon flavorings to the soy milk before fermentation begins.

Another important, perhaps far-reaching development in microbial research was reported by the U.S. Army's Natick Laboratories in Natick, Massachusetts, where scientists have isolated a special fungus which produces an interesting enzyme. The fungus uses this enzyme, a catalytic molecule, to chop up the large molecules in wood, cellulose, to produce smaller, more useful molecules of a simple sugar, glucose. This work has opened up the possibility of turning huge mountains of waste materials—such as old newspapers, wood chips, or even cow dung—into sugar. Indeed, so much sugar could be produced from these sources that it might even be economical to convert it—via fermentation, of course—into alcohols for use as a motor fuel. Experiments have already been done showing that alcohol, when mixed with gasoline, can serve as fuel to power automobiles.

Such developments mean that for industry and for genetic engineering, the future is already here. Microbes which are already widely used in industrial processes are going to be increasingly altered, tampered with, and eventually tailor-made for specific applications. Too, today's microbes, already at work producing medicines, wines, foods, chemicals, and other valuable products, will be tinkered with to improve their performance, to speed up their chemical processing, and perhaps to change the quality of the end products they create.

Even now these alterations are easy to perform on bacteria, and it's quite clear that they will be done even more, soon, on a larger scale with more ambitious goals in mind. Indeed, bacteria and fungi are probably going to become some of the most productive, most versatile "factories" in the world, pouring out vast quantities of specific and complex products of high purity at low cost.

It's interesting to remember, in addition, that some 4 billion years ago microbes were the earth's pioneer inhabitants. They remain the oldest, smallest, and most common currency in the money markets of life. Since the beginning, the microbes' role in sustaining the higher forms of life has been crucial, and now—with the advent of genetic engineering—it is, again, the microbes which are leading the way into the future, whether they know it or not.

As one scientist put it, after polishing off a long speech on the future of genetic engineering: "Perhaps someday, and not too far in the future, we'll see some gigantic new chemical company, a new du Pont or a new Monsanto, come up with this motto: 'Better Things Through Living Chemistry.'"

7

Genes: The Roots of Behavior?

PROBABLY THE last truly unexplored, unexplained frontier remaining in the life sciences is that biological enigma known as the living brain. Scientists thus far have fiddled, fussed, and probed around the edges of this uncharted territory, but the brain still represents one area where many of the most interesting and most important questions haven't even been asked yet.

Still, the brain isn't wholly unknown. Doctors have learned how to do some rather crude cutting—as in those notorious lobotomies —and neuroscientists are rapidly learning more about how a whole array of newer psychoactive drugs tangle with the brain to alter behavior. Indeed, drugs are clearly beginning to change the medical profession's understanding—and the treatment—of mental disorders.

As should be expected, genetic researchers are also showing some interest in the brain, and they're beginning to use their newest tools and techniques for gaining entry into this mass of grayish tissue, the body's master organ.

Actually, few laymen really ever take the time to read up on and begin to understand all that this three-pound blob of rubbery tissue, the brain, allows them to do. Nonetheless, the enormously complex circuitry and biochemistry of the brain are what permit vision, touch, movement, speech, hearing, memory and—best of all —creativity. Emotions, too, are part of the brain's doing, so we can blame the brain for hatred and aggression, but then also applaud it for love. But listing its functions doesn't bring us much closer to an improved understanding of the brain—learning how it's built and how it works—because the brain, as a living organ, happens to be the most complexly organized structure on earth.

The human brain does its myriad tasks while using only ten watts of power, even though its complex arrangement of wiring and its memory capacity exceed the combined abilities of a whole row of the biggest, most versatile man-made computers. In fact, during your lifetime, your brain is put to the task of gathering and storing more information than can fit inside a huge, well-stocked library. Some of the information gathered and treasured may be a bit trivial, but then libraries aren't perfect, either.

It is also known that the brain is the body's hungriest organ. Some 20 percent of the body's blood supply is always occupied with carrying nourishment and oxygen to the brain. About 1.5 pints of blood is sent flowing toward the brain every minute just to keep things running in order. And, as the master organ, the brain is so important to the body's normal functioning that in times of crisis—when blood is in short supply—the brain gets its full share regardless, leaving other parts of the body to gasp and suffer until the flow of oxygen-rich blood returns.

The brain is a formidable structure; so formidable, indeed, that before the turn of this century, few people suspected—or even hoped—it could ever be understood. Now that attitude has changed, and in the past few decades the neurosciences have been accelerating rapidly. In rather philosophical terms, the human brain is at last moving directly toward one of the most profound accomplishments ever: an understanding of itself.

Along with the growth of science in all fields, man's understanding of his own brain has changed considerably even since the 1920s and 1930s. Until then the brain was thought of as just a tangled mass of interconnected nerve tissue, growing more or less at random in response to experiences and learning. The theory then was that the brain is totally plastic; that its abilities and memory capacity are the products of what has been learned, of what has been pushed in from the outside by experience.

It was gradually discovered that the brain, as a discrete organ, is carefully put together in a distinct, orderly, well-disciplined fashion, having parts that are highly structured and patterned and which behave according to a set of master plans. These plans, of course, are the genetic instructions—the genes—which are tucked away inside each brain cell. Thus a new understanding of the brain is growing. The brain is now considered only partially as a moldable organ—changing in response to experience—and also controlled in part by its own set of genetic blueprints which seem to establish its range of possibilities and also set its limits.

Naturally enough, like all other organs of the body, the brain is built up from billions of tiny, individual cells which have specialized themselves so they can perform only their own individual tasks. Throughout the nervous system, the most important cells, in terms of action, are the neurons. But the neurons aren't alone. Altogether there are some 100 billion separate, discrete cells in the brain. About 90 percent of these are the so-called glial cells which appear, for now at least, *not* to be active in the brain's data-handling processes. Glial cells serve instead as the brain's house-keepers, carrying nutrients and oxygen from the blood to those all-important neurons. This leaves the remaining 10 billion brain cells—again, the neurons—responsible for the brain's amazing capacity to send and receive messages, store information and retrieve it, and come up with bright new ideas.

The neurons themselves appear to be among the most highly specialized cells of the body. In addition, each of these nerve cells seems to be, each in its own way, slightly different from all the other neurons, even though they may be structurally similar.

Basically, all neurons are built alike, having a main cell body which contains a nucleus holding the chromosomes, which carry the genes. These genes appear to be programmed to handle or produce some 100,000 different types of proteins.

Along with these basic internal structures, each neuron also has several—or even a dozen—tapered arms called dendrites. In turn, each dentrite can sprout hundreds of small, hairlike branches at its end, which are called the nerve endings. These nerve endings are the terminals, the antennae or receivers which collect signals from the other nerve cells.

In addition to these structures, each neuron also has a long, thin tail called an axon which is also tipped with many branches. The axon is the transmitter "wire" through which the nerve cell sends its own special signals to other cells. In most instances, these long axons are tiny—even microscopically small—but there are a few—such as those which send commands far away to muscles in the legs—which can be five feet long.

Very few of these nerve cells actually connect physically to each other. Between any single dendrite and the cell it communicates with is a microscopically small gap, the synapse. The messages flow across the synapses from cell to cell. And because of the large number of branching dendrites and their numerous nerve endings, each cell has the potential for making an enormous number of connections with other cells. For example, the cerebral cortex—the portion of the brain responsible for speech, for the emotions and thinking—has cells which are known to have 80,000 or more synapses—or communication links—with other cells. Each of these cells can be bombarded constantly with messages from other cells and can transmit its own response—a similar message—out through its own axon.

Brain researchers have finally come to suspect that each individual nerve cell is constantly bombarded both with positive and negative electrical signals coming from the other nerve cells. When a sufficient charge is built up in one direction, the assaulted cell fires its own message, which joins the millions of other signals racing around in the brain's complex circuitry.

We now know that these signals jump across those small synaptic gaps from cell to cell through chemical transmission, by the sudden release of ionized substances such as sodium, chlorine or potassium. These ions permeate the cell membrane at the synaptic tips and open up the gates to allow passage of a signal. Despite such findings, however, the complex biochemistry of signal transmission—and the control of this biochemistry through genetics—is still not well understood.

With all of this in mind, then, it's apparent that the brain still offers a complex and challenging new territory for scientific exploration; and the specialists now working to perfect the tools of genetic engineering are already finding some useful and interesting techniques. This approach is possible because the genes a creature inherits from its parents largely determine how the creature is built and controlled. Thus the genes are in charge of the nervous system's basic architecture, commanding its internal wiring, its chemistry and its pattern of growth. Given this important fact, then, one would expect that the genes are also a controlling factor in an individual's behavior, in its basic intelligence and in its responses to stimuli.

Such an assumption, however, will draw the unwary into one of science's longest-standing, most unresolvable squabbles. Indeed, if there's anything resembling an old-fashioned barroom brawl in the somewhat stuffy halls of academe, it has to be this fight over behavior, this bitter old argument over nature vs. nurture. It has become almost a knock-down, drag-out affair, with experts and pseudo-experts on both sides calling names, challenging each other's honesty and intellectual capacity, arguing that their own views represent True Gospel—if there is such.

Most of this continuing sound and fury now revolves around problems of race and racial prejudice, especially as the argument over nature-nurture is seen in the universities. Even though geneticists have generally succeeded in discrediting older racist ideas that one population group is inherently genetically better than another, a few scientists are still prepared to argue that some races—and especially blacks—are inferior intellectually and that such differences can be explained genetically. They claim, too, that they aren't being racist, stating mainly that the research to answer this question once and for all should be done.

Opponents—perhaps too hysterically—remain adamant that such research shouldn't even be thought of. Actually, modern genetics shows rather clearly that the world's different racial stocks are now so thoroughly mixed that no pure races exist. There is also strong evidence that the differences between races which show up through intelligence tests are probably more a result of different cultural influences than levels of intelligence. Apparently, no easily provable—or obvious—superiority is granted to one race over another by nature, so the answer seems to be that nurture—

meaning good prenatal care, birth in a clean hospital, adequate nourishment, a stable family environment, and an intellectually stimulating home—plays an important—perhaps dominant—role in the results that emerge from intelligence tests.

Unfortunately, what seems clear is that whatever research is done, and however the results come out, this argument over the sources of human behavior won't be resolved soon. Too many stubborn individuals have too much emotional energy invested in the squabble to give up easily. Worse, most of the experiments designed to answer whether it's man's genes or his environment which determines why he acts as he does can usually be interpreted in a hundred different ways—and often are. Basically, the argument still surrounds this one simple question: does the human animal behave in his own perverse manner just because of the cards—the genes—he was dealt, or because of the game—the environment—he finds himself in?

The answer to both sides of this fuss, it seems, is yes. At best, the complex actions and reactions known as human behavior can be seen as an unholy mixture: bits and pieces of experience, a few inborn inclinations, and a whole pot of circumstance, all thrown together into a simmering behavioral stew. Sorting out the genetic meat from the environmental potatoes—without discarding the rich human gravy—seems an impossible task.

Certainly nobody yet has the answers in this fierce fight and, perhaps in the end, nobody will care because in some ways it seems to be particularly a fight between academic ivory towers, with verbal broadsides being hurled back and forth between geneticists, sociobiologists, other scientists and a few nonuniversity-based outsiders. Some observers, indeed, are beginning to find the whole brouhaha rather boring.

Nevertheless, and despite all the arguing, as the work in the brain sciences has been carried forward, biologists and other scientists have been gathering more small and important bits of information about how the brain gets built. They've long known that the human brain is basically an expanded organ that has developed gradually, layer on layer, from a series of similar, less complex structures over millions and millions of years of slow evolution. By watching the orderly development of embryos, they've also discovered the natural order in which parts of the brain are laid down as the individual grows inside the womb. They've been able to keep watch on the long, wirelike nerve fibers reaching out purposefully to make just the right connections. One of the puzzles still remaining is how each fiber "knows" it is supposed to wire itself up to a certain spot. It is generally agreed now that the genes somehow direct this process, but there is no agreement yet on exactly how each neuron knows which connection—and only that connection—to make.

Worse, what also isn't known, among other things, is:

First, what is memory? Is the memory process a series of chemical changes which occur inside certain neurons? Is memory an arrangement of electrochemical linkages that can be changed, like changing the wiring or the arrangement of switches in a computer? Or does memory involve the production of new proteins inside the brain cells in response to experience? Could memory be some combination of all of these?

No definitive answers are yet near at hand, but in recent years advances in neurochemical techniques have allowed researchers to study some of the small chemical events that take place when an individual learns something. A quick electric shock, given almost instantly after something has been learned, can apparently erase a trace of memory. And, even though there is no agreement yet on what the precise learning process is, many scientists are beginning to suspect that memory—or learning—involves the establishment of new nerve pathways or the reinforcing of older pathways. The central idea seems to be that the formation of new pathways is made possible by chemical events in the brain and that this involves building some new copies of that familiar genetic messenger molecule, RNA (ribonucleic acid).

One group of researchers in England—Steven Rose, Pat Bateson, and Gabriel Horn—began investigating the chemical changes—specifically, the production of new RNA molecules which occurs in the baby chicken's brain as it is imprinted soon after hatching. It is well known that many birds imprint—or learn to follow their mother—soon after they first see her movements, usually within a few days of hatching. It has also been found that chicks deprived of their mothers can be imprinted on other moving objects—or even on flashing lights.

For purposes of the experiments done in England, this research group injected radioactively labeled precursor molecules for RNA into the chicks soon after they had gone through the imprinting experience. Later, analysis of how much of the radioactive material had been built into the brain's RNA supply provides a crude measurement of how fast RNA was made during the learning process. Results of these measurements suggest that learning stimulates increased RNA synthesis inside the brain cells, and thus that learning involves the production of new proteins.

Unfortunately, research on the learning process hasn't gone far enough yet to explain how production of protein works in the memory process, or how it might lead to altered nerve pathways during the learning process. Still, some scientists are beginning to favor the idea that signal-carrying nerve pathways may be altered rather easily simply by changing the ability of individual nerve cells to pass information around among themselves. This change might occur through some sort of electrical or chemical change at the synapses, those little gaps separating nerve endings which serve as the transmission sites between nerves cells. A group of research specialists at the University of Illinois has found some evidence through anatomical studies that nerve pathways are indeed altered by learning. Their work with laboratory rats indicates that after the rats have undergone several weeks of training, the nerve cells in the brain's cortex showed more branching than similar nerve cells in the brains of their untrained litter-mates. These litter-mates—serving as controls—were handled the same way, but without training.

As for the methods discovered for blocking memory, in addition to the use of electric shock, Bernard Agranoff, a researcher at the University of Michigan, found that a drug called puromycin, which inhibits the production of proteins, can effectively block memory storage in goldfish. Still, there is one catch. The drug must be administered shortly after the learning occurs; otherwise the memory takes hold and stays. Indeed, a goldfish injected with puromycin one hour after being trained did not forget the task it had been taught. The lesson here—for the scientist, at least—is that such protein-synthesis-blocking drugs must be given early—almost immediately—to be really effective. A second implication is that puromycin is able to block the formation of the memory molecules, but cannot unmake the protein once it has been manufactured. Other chemicals, however, may be found which are able to do this, which are able to erase long-term memory even though it has been "set."

One of the factors that seems to be important in how "set" a memory is involves the amount of training the creature went through. A very well-trained animal generally needs a much larger dose of the drug to block formation of a memory than does an identical animal which has only poorly learned its task. This would suggest that the well-trained animal was making more of the special memory chemical, probably a protein, than his less-trained companion.

Second among those puzzles about the brain is the question of why there are two kinds of memory: short-term and long-term? Also, what is the seat of short-term memory, and where is long-term memory stored, and in what form?

Third, how are memories that are stored in the brain recalled so quickly and accurately when needed? Are they perhaps stored as chemical molecules that can be unrolled and read like a scroll or a tape recording? And, perhaps more important, how do these memories sometimes interact and bounce off each other to spawn new ideas? Indeed, what is the process of creativity? What is imagination?

Fourth, what is the connection, if any, between brain diseases, nervous disorders, mental illness, and the genes? Is society wasting time and money sending its mentally ill patients to psychiatrists when they should be treated instead with drugs? Or should drug therapy be ended in favor of psychiatry?

Of course, these aren't trivial or easily answered questions. But it illustrates how wide-open the whole subject of brain chemistry is. Scientists are still finding new chemicals in the brain that they didn't know existed. It's well known, for example, that one of the body's most important ways for controlling itself is through the use of chemical messengers called hormones, a good example of which is the "fight-or-flight" hormone, adrenaline. Such hormones are carefully built according to the precise instructions in the genes, but they are released at the proper time on instructions from the brain.

Newer findings in brain and hormone chemistry are also hinting that disorders like schizophrenia and possibly as many as 150 other mental problems—such as the manic-depressive syndrome— may be the direct result of chemical or hormonal imbalances. Some recent findings suggest that schizophrenia may be a meta- bolic disease, at least to some extent, since it sometimes appears to be associated with a nutritional abnormality which strikes in early childhood, celiac disease. Basically, celiac disease involves the digestive system's inability to handle the portion of wheat flour known as gluten.

Also under scrutiny as a possible problem of hormone imbalance is homosexuality, for which evidence is still accumulating to sug- gest that homosexual men and women often have abnormal amounts of one sex hormone or the other in their blood. What this suggests is that an excess—or perhaps a deficiency—of a sex hor- mone may lie at the root of homosexuality or, conversely, it could mean that homosexuality might itself stimulate hormonal im- balances. As with so many other medical problems, it will prob- ably be difficult to sort out the cause from the symptoms. Research is continuing.

Some of the ideas that researchers come up with for explaining behavior through biochemistry occasionally get shot down, too. It has been found that people born with abnormalities of the sex chromosomes—the XYY or XXY syndromes—are *not* particularly violence-prone. The idea that the presence of an extra sex chromo- some signals a propensity for violence has now been almost com- pletely discredited as research has continued. Studies now indicate that persons carrying extra sex chromosomes have no more vio- lence in their lives than normal members of the population; but there may be a tendency to be less capable mentally, of lower IQ.

Obviously, then, such examples of mistakes seem to lend a

sense or urgency to the study of genetics and how the genes relate to the brain and its function. As research on chemical imbalances and their impact on the nervous system continues, more and more excitement is being generated as discoveries—one by one—suggest that abnormal brain chemistry may lie behind many mental illnesses. The importance of this is that now, for the first time, physicians are looking toward possibly effective ways for finally curing—or at least managing—some serious mental disorders. Schizophrenia is just one of many that may finally be deciphered.

At an international congress on biochemistry in 1976, the Salk Institute's Dr. Roger Guillemin proposed, as a working hypothesis, that his group may now have found one biochemical mechanism that is involved in mental diseases. What he reported, from a series of animal studies, was isolation of brain substances which produced symptoms which resemble schizophrenic catatonia in human beings. Guillemin said he thinks it will be easy to determine whether some connection between the chemicals and the disease exists.

Guillemin presented laboratory results showing that several small brain chemicals, certain peptides, may be closely linked to definite types of brain function. Three of these chemicals were isolated, purified, and later synthesized by Guillemin's research group in La Jolla, California. These three peptides have been named endorphins, and the La Jolla group reported that they act either as tranquilizers, or they produce violent behavior or cause prolonged catatonia when injected into laboratory rats.

Guillemin's work on such elusive chemicals is an outgrowth of work done earlier by Dr. John Hughes in Aberdeen, Scotland. Hughes's work indicated that the brain might use such small chemicals as natural pain killers, with the tiny molecules binding—or sticking to—the receptor sites on brain cells, thereby blocking transmission of some signals.

Guillemin, who had been working in a similar area, had discovered earlier that the brain's hypothalamic area naturally produces a class of peptides known as hormone-releasing factors. These small chemicals tell the body's master gland—the pituitary

—to either stimulate or inhibit the release of various hormones such as insulin, glucagon, or somatostatin.

In their later experiments at the Salk Institute, Guillemin and colleagues began injecting small doses of these three new peptides—the endorphins—into rats and found that each chemical produces its own typical, specific reaction. One of the endorphins produced a mild tranquilizing or pain-killing effect. The second aroused the rats to extremely violent behavior. The third, which Guillemin said was by far the most active, produced a catatonic state which lasted more than three hours. During this period, the rats remained rigid. To end this freeze the rats were given a morphine antagonist—naloxone—and within a few seconds those once-stiff rats were up and running around.

Similar work with other interesting brain chemicals now seems to be opening up a wide array of psychoactive substances in addition to the endorphins reported by Guillemin. One research group at the National Institute of Mental Health has found another natural pain-killing substance which circulates in human blood. It acts as a morphinelike sedative and has been named anodynin.

Several years earlier a research scientist in New Orleans, Louisiana, reported finding another brain substance that is both interesting and controversial. This chemical—now known as scotophobin—is claimed to code for one specific behavior: fear of the dark.

In this research, the technique involves grinding up the brains of the trained rats, then injecting a brain extract into untrained rats who, under the right circumstances, then begin exhibiting an unnatural fear of the dark. This work has also been extended to goldfish, and appears to produce similar results. But the work is still very controversial, and many scientists are inclined not to believe it—or at least to withhold judgment.

Not quite so controversial, but no less interesting, is the work being done by separate research groups in Boston, Massachusetts, and Tokyo, Japan. Both groups have found that the living brain manufactures what seems to be its own specific chemical that is now being referred to as a "sleep factor." The chemical apparently accumulates during wakeful hours, then gets used up and disappears during sleep. This substance also appears to be a small, simple protein—again a peptide—which apparently exhibits the same activity in man, mouse, sheep, goat, and cat.

The Boston group, working in Dr. John Pappenheimer's laboratory at the Harvard Medical School, has been able to show that this substance—call it drowsy syrup—found in the spinal fluid of goats is able to promote sleep in other animals. In Japan the workers reported they have also extracted and purified a sleep-inducing chemical from the brains of 1,000 sleep-deprived rats. The substances isolated by the Tokyo and Boston groups may be identical. Most recently, the Boston group found that their "drowsy syrup," taken from human spinal fluid, is active when injected into goats.

Since such brain chemicals appear to be natural products, it would suggest they are manufactured by special cells and are thus coded for by the genes. This would lead one to suspect that different gene combinations—or even some gene mutations—could account for why some people are more wakeful or drowsy than others. Better yet, discovery of such small, powerfully active molecules encourages the possibility that scientists, someday, may be able to duplicate this natural "drowsy syrup" for use as a new type of sleeping potion, a potion that will induce a natural form of sleep without depressing the central nervous system so drastically.

One important target of such work, certainly, will be schizophrenia. As mentioned earlier, there are some strong hints now that this form—or forms—of mental illness may be linked to hormone, chemical, or metabolic imbalances. One research group, headed by Dr. Charles Frohman at Wayne State University in Michigan, has already reported isolation of a strange new protein from the blood of schizophrenic patients. At first Frohman's group suspected that this new substance would be found only in the blood of mental patients but not in the blood of normal people— at least not in the same amounts. Instead, however, tests indicated that the chemical is found in roughly equal amounts both in patients and in nonpatients.

This was a setback, but after returning to the laboratory to rethink the experiments, it was soon found that this mysterious chemical comes in two forms, and that schizophrenic patients seem to be carrying an abnormally high amount of the unusual form of the protein. Frohman's group found that in schizophrenia patients some 75 percent of this one chemical was made up of the odd-shaped molecules, described as resembling an elongated basketball. Normal people, carrying equal amounts of the substance, carried only 10 percent of it in the odd shape. In its more normal form the protein occurs as a long, narrow molecule which tends to flow freely. In both molecules, the basic chain of amino acids is thought to be identical. The only difference seems to be that the abnormal molecules, for some unknown reason, fold up differently and thus react differently. And one strong possibility is that some unknown genetic mechanism lies behind this abnormal folding of the protein molecules.

Such findings show that natural chemicals play an important role in how the brain functions normally, and they're also showing ways by which the brain might easily be led astray by introduction of powerful drugs, or by unnatural forms of natural substances. But what about the genes? What role do they really play, especially in abnormal states?

At present, these questions are mostly unanswerable, but scientists, when faced with problems of complexity, begin looking for ways to simplify. In other words, when the human brain presents too many complicated barriers which block understanding, one powerful approach is to look lower on the evolutionary scale for simpler systems which might yield important clues about some of the basic rules underlying all living nervous systems. The best example of how simplification can pay off came years ago in the early study of genetics and biochemistry, studies which eventually turned into the productive and exciting field of molecular biology. Scientists like Salvador Luria and Max Delbruck, now at the Massachusetts Institute of Technology and the California Institute of Technology, respectively, began looking at tiny creatures called "phages"—bacteriophages—which are viruses that make their living by infecting bacteria. Phages are simple enough organisms so they're potentially understandable, and the use of an understandable system to ask important questions led to some of the best answers to come out of biology. Indeed, the use of phages has helped scientists unlock some of the most basic, important secrets of the living world.

Similarly, the neuroscientists have been seeking simpler, more understandable nervous systems to study in hopes of learning some of the important secrets hidden in the complexity of the brain. One such find was in the squid, which has a giant axon that is easy to find, isolate, and study. Studies of nerve-signal conduction, the chemistry of the synapses, and other experiments have been possible—and the results understandable—because of simplification.

Sometimes, however, once the maximum information has been squeezed from a simple system, it's necessary to move to a more complex organism in the search for answers. This is now occurring in brain studies, especially as scientists keep asking questions about the genes and how they relate to behavior.

Those are the questions which remain unanswered, but Dr. Seymour Benzer, also at the California Institute of Technology, and his colleagues at some other research institutions are uncovering some fascinating clues about the genetic basis of behavior by manipulating the genes and the cells of living fruit flies. Obviously it's a long, even perilous leap up the evolutionary ladder to equate the behavioral abnormalities of the fruit fly with human behavioral problems, but the temptation to make that leap is often present when the results of research stand out so sharply. Too, even if the answers Benzer is finding are not directly applicable to the human condition, they do seem to be leading in the right direction.

Benzer, who first made his mark in science as a solid-state physicist, later switched to biology and began dissecting behavior in the fruit fly by causing mutations of the genes. His objective, basically, has been to discover what part the genes themselves play in behavior, to match certain types of behavior with certain genes, and then to find out exactly how and where the gene works, and how a malfunctioning gene influences behavior. According to an article Benzer wrote for *Scientific American* magazine, his basic approach is to try to keep the fly's environment constant, change the genes, and see what happens to behavior.

Benzer has also noted that he chose the lowly fruit fly—already a strong favorite among geneticists for use as a research tool— because *Drosophila melanogaster* represents a compromise. In terms of its mass, in the number of neurons in its nervous system, and because of its rather fast breeding rate, the fruit fly stands about halfway between *E. coli* and man. The bacterium, certainly, would represent probably the simplest type of nervous system, the single neuron. The fruit fly, however, like the human, has a substantial number of neurons, lots of dendrites, axons, and synapses, plus an array of chemical molecules that act as transmitters or message carriers.

The basic assumption behind such work, of course, is that the genes *do* have something to say about behavior; at least in how the various structures of the body are built and maintained.

Indeed, early experiments by Theodosius Dobzhansky and Jerry Hirsch indicated that, through careful, selective breeding, it is possible for various types of behavior to be enhanced if the program is carried on over enough generations. Even alone, this result shows there is some genetic component to behavior, and it suggests that behavior can indeed be modified through genetic manipulation. Unfortunately, in this sense behavior appears to be under the control of a large number of genes, so it remains extremely difficult to sort out individual behavioral effects that might be produced by single genes.

Nonetheless, Benzer and his coworkers have been able to dig out a few interesting and revealing behavioral abnormalities by causing mutations in fruit fly genes. To accomplish this, a population of fruit flies is exposed to radiation or to chemical mutagens, both of which can cause gene mutations without killing the flies. The mutations often show up as behavioral abnormalities, but they can also be seen, or marked, by producing obvious changes such as white eyes or forked bristles.

Along with such obvious markers, Benzer and his colleagues have found—in some cases—some behavioral abnormalities that are strikingly similar to behavioral problems seen in human beings. In most cases, for purposes of identification, these odd flies are named according to their particular disabilities. For example:

• *Easily Shocked:* When subjected to a mechanical shock, such as their jar "home" being slammed on a table top, flies with this mutation fall into what resembles an epileptic seizure, lying on their backs, flailing their legs and wings, coiling abdomens under, and passing out. After a while they wake up, stand up, and move about as if they were normal. Another researcher, Burke H. Judd at the University of Texas, has also produced this mutant, but he has named it TKO.

• *Paralyzed:* This mutant, discovered by David Suzuki and his colleagues at the University of British Columbia, is temperature sensitive, collapsing when the thermometer moves above 82° Fahrenheit. When the temperature comes down again, *Paralyzed* gets up and moves about as if nothing had happened. *Comatose,* another version of this mutant, takes longer to recover—from minutes to hours—depending on how long it has been exposed to high temperature.

• *Stuck:* This fly suffers from an embarrassing inability to disengage at the end of the normal twenty-minute copulation period.

• *Coitus Interruptus:* As suggested by its name, this fly suffers from the opposite problem; automatically disengaging halfway through the normal copulation period, never producing any offspring from such incomplete matings.

• *Hyperkinetic:* Grossly overactive, this fly consumes oxygen at an ultrahigh rate, lives fast, and dies young.

• *Nonclimbing:* Unlike normal fruit flies, which have a propensity to climb upward on walls, this fly is content to stay on the flat and level, climbing nowhere.

• *Flightless:* Even though its wings are perfectly developed, and even though the male can vibrate his wings in the normal courtship ritual, these flies can't—or won't—fly. When dropped off the end of a glass rod, they fall like stones.

• *Shot-Full-of-Holes:* This fly lives a normal life into adulthood, then suddenly begins losing control, staggers around, and falls over dead. On examination, its brain is found to have thoroughly deteriorated, being riddled with holes. This syndrome seems to be somewhat analogous to Huntington's chorea, a disease in humans.

Such mutant flies are interesting, of course, but Benzer cautions that even if it is known that a certain mutation causes a particular kind of altered behavior, this knowledge still doesn't lead a scientist directly to the spot in the body where the gene has caused the problem. He notes, for example, that in some cases in humans, in which the retina of the eye degenerates, there is nothing actually wrong originally with the structure of the eye or with its abilities. The problem instead is in the person's intestine which, for some reason, is unable to absorb vitamin A from the food. Lack of vitamin A, then, leads to degeneration of the retina.

To help untangle such genetic and cellular puzzles, Benzer, his coworkers, and researchers in other laboratories have been building what they call mosaic flies in which some tissues are normal while others carry one or more mutations. It's possible to build such strange creatures because scientists have located some flies which have an abnormal sex chromosome, a ring-shaped X chromosome. In addition, this abnormal chromosome has a tendency to "fall out" or disappear from some cells during the division process. Thus a fly can end up with two kinds of cells, with some cells that were originally female cells (with two X chromosomes) turning abruptly into male cells when that chromosome is dropped. The rest of the cells, still carrying two X chromosomes, remain female. The result, then, is production of a fly which is half-male and half-female. Which part of the fly remains female or becomes male depends on the original location and orientation of the odd cells at the time the ring-shaped X chromosome fell out.

In research work with such flies, the trick is to find flies in which behavioral mutations have occurred on one of those sex chromosomes. So, when one of these chromosomes has dropped out, the fly emerges not only half-male and half-female, but one of those sides, or halves, also carries tissues which have the mutant trait. In addition, it's also possible to "mark" the abnormal tissues—at least on the fly's exterior—by including a mutation for odd color, or white eyes or forked bristles, making the mutant parts easy to see.

An interesting point, too, is that the dividing line between these halves, between male and female cells in a single fly, can run in any direction, such as down the midline so that one side is male, the other female, or at the neck so the head is male and the body is female. Or the fly can emerge divided diagonally.

Such tricks might sound like weird things to do to undeserving fruit flies, but for scientists these techniques present useful ways to compare the effects of normal genes against the actions of abnormal genes, all in one animal. Better yet, it's a method that can lead to finding exactly where a mutation exerts its primary effect. Benzer reports one example in which a fly known as *Wings Up* walks around full time with its wings poking straight up instead of laying them flat against its back like normal flies. Analysis of a mosaic fly with this *Wings Up* trait led to discovery that the genetic defect occurs inside the thorax. Even though the wings seem to be the abnormal part of the fly, it is the muscles inside the thorax which cause the wings to lock in the up position.

The importance of this work is that it represents a strong start toward sorting out the complex relationships that seem to exist between the genes and an individual's behavior. Indeed, Benzer's work shows that important behavioral changes can certainly be wrought through mutation, especially when a mutation causes a physical defect which makes the individual less able to perform normally, less able to cope with its normal environment.

While trying to keep all this in mind, we should also look for developments in the behavioral sciences that begin to detail how the brain works, how it helps each individual cope with its environment. As the sciences are progressing now, it shouldn't be long before a much better, clearer understanding of the brain is reached.

As for the potential of efforts to exchange genes, hoping to alter behavior—the real stuff of genetic engineering—it can be expected someday that this will be tried, aimed possibly at improving the brain itself. New genes, dumped in with the hope that the brain will emerge somehow bigger or better, will probably be added to animals first to see how it works. Then, years and years in the future, it may be tried on man—if there's any potential for benefit.

Before then, however, there is already reason to believe that it's

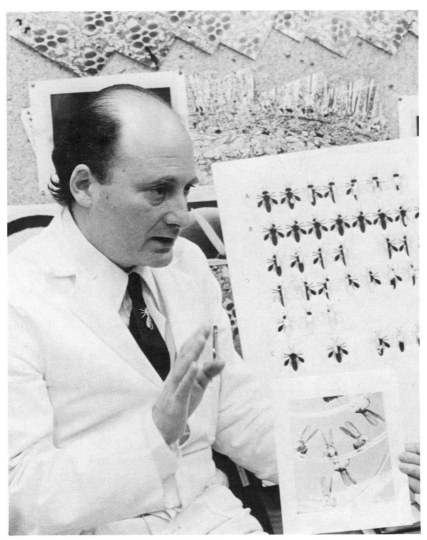

Dr. Seymour Benzer, in his office at the California Institute of Technology, discusses the role that some genes play in controlling behavioral quirks in the fruit fly. By deliberately causing mutations—or changes—in the genes, Benzer and his colleagues can alter the behavior of different fruit flies in strange ways. Through other techniques, they can also learn how specific mutations alter the fly's normal structure to produce behavioral changes.

already possible—and perhaps reasonably simple—to cause an increase in the number of neurons in the human brain. There are some strong hints, coming from research with rats, that by merely giving an unborn infant better nutrition—indeed, supernutrition—it is possible to significantly increase the basic number of neurons in the brain. There have been some reports, too, that the rats so treated have emerged from the womb and have grown up to have better mental capabilities than normal rats.

Watch, too, for efforts someday to find ways to preserve the number of neurons people have naturally, keeping them from dying off regularly and rapidly as a person ages. At present, most humans begin losing thousands of neurons daily at about age thirty-five. This is a natural process, and the neurons are never replaced. This one fact, indeed, says biologist James Bonner, is what mitigates now against heroic efforts being made to prolong the human lifespan. He suggests that if the brain is gone, or nearly gone, there's not much point in preserving the living body if it's just an empty brain case.

This suggests that until there's some real, honest progress toward protecting the brain from the forces of natural decline—or toward somehow resupplying the brain with a fresh batch of lively neurons—there's little point in performing lifesaving heart, lung, and liver transplants except in special cases. It also suggests that all the tricks being dreamed up for use in the field of genetic engineering, especially those aimed at stretching out the lifespan, may be useless unless some are used to extend human abilities and prolong the usefulness of the brain.

8

Hazards: New Faustian Bargain?

PERHAPS, NOW in retrospect, all the fear and trembling was a bit premature, but when American astronauts brought their first batch of precious rocks and soil home from the moon, that load of minerals was treated with more caution and more respect than some of the really deadly diseases already known to exist here on earth. Extreme caution seemed worthwhile, even if it was only in response to a list of questions which all began with "What if." For example:

"What if the moon, sometime in the dark, mysterious past, was once a comfortable habitat for living organisms?

"What if, even now, in the searing heat and numbing cold of the moon's surface environment, something alive still slumbers?

"What if man's rock-collecting astronauts bring back a few living microbes either with their rock and soil samples, or on themselves?

"What if, once these microbes awaken in the earth's rich atmosphere, they begin multiplying, spreading like wildfire to set up new colonies, warm and comfortable in their new home?

"And what if these strange moonbugs are deadly to man?"

Obviously, and fortunately, none of these hypothetical questions worked out in practice, but at the time, back in 1969, the possibilities were taken seriously enough so that ultrasecure laboratories were set up to handle the Moon samples, and the astronauts themselves were locked up in rigid quarantine for weeks.

The point, of course, is that similar precautions might now seem to be in order for an equally threatening list of "what if" questions that are being raised by the emerging science of genetic engineering. These questions might not seem as obvious, especially when compared to the adventure and the mystery involved with the possibility of moonbugs, but there is the prospect that some scientists, now tinkering with the genes of germs will inadvertently—or even purposely—create strange new organisms that are more dangerous and more damaging to human health than anything the moon might have spawned.

And some of the work in that direction is already being done under less-than-secure conditions, increasing the real potential for launching a quiet, creeping disaster. Indeed, as things are going now, it seems more and more possible we'll be seeing a home-grown "Andromeda strain" in real life, perhaps running loose in our biggest cities.

A city that got a strong taste of what creation of a strange new disease might mean was Philadelphia where, in the summer of 1976, an outbreak of something now called "legionnaires' disease" killed 29 people and briefly disabled 151 more. Like a new, creeping plague of unknown origin, legionnaires' disease hit, killed, then left no footprints, no clues, no calling cards.

Epidemiologists were stumped. Was it the dreaded swine flu? Viral pneumonia? Toxic chemicals? Pollution? What?

No obvious answers emerged, and Philadelphia became, in the minds of some scientists and observers, a frightening model for what could happen, perhaps even on a larger scale, with the escape of a newly manufactured, unstoppable bug.

Even so, for most people, the full potential in the new tricks of genetic tinkering remains obscure, an undecipherable riddle—probably because they haven't been paying attention. Nonetheless, there's very little doubt now that mankind is teetering right on the brink of a true genetic revolution, or one might even call it a Second Genesis. What it's amounting to is another of *Homo sapiens*'s brave, brash—and perhaps foolish—attempts at playing God. Already, having been given the tools, the talent and the time, scientists have nudged the craft of genetic manipulation off that shelf reserved for science fiction. Now they've rolled it out into the sunshine, into the bright world of experiment and application.

It may not be obvious, but the changes wrought through genetic experiments promise eventually to alter human society at least as much as, and probably more than, the introduction of atomic power, motor vehicles, and worldwide communications. The new elite which will emerge during this profound alteration of natural conditions will be the scientists, especially the high priests of biology and biochemistry who have already donned the robes of authority, who have begun playing lordly games with the most basic elements of life, the genes.

The really awesome, unsettling part of this whole revolution, however, is that it can run away, that it can escape and keep right on going. Indeed, some scientists are warning that their colleagues doing the laboratory work really have very little idea of what the results might be; what it is they're really building. The first, most basic cause for worry, even among the bioscientists, is that some fresh new disease organism—such as a modern version of bubonic plague—will find its way out of someone's laboratory and begin infecting huge populations. And the list of those frightful "what if" questions goes on and on, making those questions about the moon seem simple, if not trivial.

Nonetheless, the work continues. Even though solid results—such as those newly designed types of plants, animals, and insects—aren't yet in hand, what can already be seen is that genetic engineering is fast becoming an important social and political issue. That long list of "what if" questions is already being thrashed out among bioscientists, and the argument is obviously going to be debated—perhaps with more emotion than knowledge—among doctors, economists, housewives, students, and just about everyone else.

As might be expected, much of the early squabbling in scientific circles dealt with the perceived threats to academic freedom, mixed liberally with the fear of possible government intervention into biological research. But in the background, too, was that nagging suspicion, that persistent fear of what genetic engineering itself might do, even accidentally.

Obviously, many scientists are still unconvinced that any form of regulation, especially from government, is necessary. They fret, too, over who will be in control and how control will be exercised. But, as with atomic power, with automobiles and with communications systems, some forms of control are going to be imposed; indeed, they're already being imposed. The machinery is already in motion, and it's starting at both the local level and at the federal level.

As is true for telephones, which bring both good news and obscene calls, with automobiles which provide transportation but also run over people, and with atomic energy which produces electric power but also works well in bombs, genetic engineering can be plainly seen to be offering great potential for easing some of mankind's most stubborn burdens, balanced with the potential for harm. Like these other phenomena of modern life, genetic engineering is going to be something of a double-edged sword, a sword which—during some of the early cutting—will certainly draw some blood from both sides. We may see the benefits, but the mistakes will be there to accompany them. And the question that most needs answering: can we survive the mistakes?

One should not assume, of course, that the world's scientists are bent on doing harm. Nothing malicious is involved; there's no intent to maim, kill, or even scare. To be fair, the scientists who invented the new genetic manipulation techniques, and those who are using them, are the ones who actually spoke out first; who sounded the first alarms and became obviously worried about the unexpected, undefined hazards involved in mixing and matching batches of unrelated genes. Thus it was the biologists, the biochemists, and the geneticists closest to this work who began assessing the possible consequences of uncontrolled genetic tinkering.

What the scientists fear most is either the deliberate or accidental creation of new living organisms—especially dangerous forms of viruses or bacteria—which might escape from the cozy confines of a laboratory and set up deadly housekeeping in an unprotected world. The possibility of some lethal organism escaping from what is presumably a safe, secure laboratory, isn't as unreasonable as one might think. In England, for example, two women recently died as a result of one worker—accidentally contaminated with smallpox virus—carrying the disease outside the laboratory. In this one case, however, it was fortunate indeed that most people in Western nations are still immunized against smallpox. This one example does show, however, how easily some unrecognized, lethal new organism might find its way out of someone's laboratory and into an unprotected population.

In the United States, too, the nation's most ultrasecure laboratory, at Fort Detrick, Maryland, was unable to completely stop the movement of dangerous bugs even among the most careful and most professional experimenters. The extracautious "containment" strategies at this, the U.S. Army's biological warfare laboratory, were unable to prevent 423 cases of accidental infection by disease organisms—and 3 deaths—in a period of twenty-five years. This would at least hint, then, that reliance on containment of viruses and bacteria within physical barriers won't be the whole answer.

So far, what all these unsettling questions about biological safety have done is cause an old-fashioned family fight among scientists. On the one side are those yelling, "Whoa, let's look at what we're doing before diving all the way in!" On the other side are the anxious workers who respond, "Hey, don't hold me back. I've got some good ideas—ideas that may save some lives—and I want to pursue them before somebody else does. Besides, who are you to take away my freedom of inquiry?"

One might suspect, too, that one part of this urge to pursue research is associated with that ego-feeding malady known as Nobelitis—the possibility that this creative work will lead, almost inevitably, to Nobel Prizes. Even if prizes don't result, however, this work is going to lead to enormous prestige for scientists who make the key discoveries. And this means that research in the life sciences has now become something of a horse race.

There are, nonetheless, some biologists who are strongly advocating that the research in what is officially known as "re-combinant DNA" be stopped, or at least slowed or postponed. Prominent among them is Dr. Erwin Chargaff of Columbia University, in New York City, who in a letter to the editor of *Science* magazine declared:

> The principal question to be answered is whether we have the right to put an additional fearful load on the generations that are not yet born.
>
> Our time is cursed with the necessity for feeble men, masquerading as experts, to make enormously far-reaching decisions. Is there anything more far-reaching than the creation of new forms of life?

Chargaff's main argument, which has also been voiced strongly by other scientists, is against the use of one very common type of bacterium, *Escherichia coli,* a normal, ubiquitous inhabitant of the human gut. Known commonly as *E. coli,* this organism is found almost everywhere on earth, and especially wherever man has gone. The reason biologists like to use *E. coli* as the receptacle for strange new genes is that this organism is by far the best studied of all the bacteria and, as a favorite laboratory tool, is easier to manipulate. With this in mind, combined with the fact that *E. coli* is comfortably at home in the human intestine, Chargaff declared:

> If our time feels called upon to create new forms of living cells—forms that the world has presumably never seen since its onset—why choose a microbe that has cohabitated, more or less happily, with us for a very long time?
>
> If Dr. Frankenstein must go on producing his little biological monsters—and I deny the urgency and even the compulsion—why pick *E. coli* as the womb?
>
> Who knows what is really being implanted into the DNA of the plasmids which the bacillus will continue multiplying to the end of time? And it will eventually get into human beings and animals despite all the precautions of containment. What is inside will be outside.

Chargaff also asked strongly, "Have we the right to counteract, irreversibly, the evolutionary wisdom of millions of years in order to satisfy the ambition and curiosity of a few scientists?"

The efforts to awaken the biology community—and later the public—to the possible hazards of this exciting new line of research began small, and most agree that they were first voiced in the laboratory run by Dr. Paul Berg at Stanford University.

Berg and his colleagues had just achieved success with their new techniques for snipping and recombining small rings of DNA, and they were finally learning how to isolate small snippets of this genetic material and hook them together in unique new combinations. Later, during what has become famous as the Asilomar Conference, Berg recalled that, in his laboratory at Stanford, he had been getting telephone calls almost daily.

"What do you want to do?" Berg said he asked other scientists when they requested samples of his new plasmid, called pSC101. Berg said his callers sometimes described what he called "some kind of horror experiment" and that, on further questioning, it turned out they hadn't thought about the consequences of such experiments. Berg noted, too, that he couldn't criticize these callers because he also, some two years earlier, had been in the same position.

Berg's growing apprehension about the potential dangers of linking strange, unrelated bits of DNA together was finally shared more widely in meetings with other bioscientists in 1973. By July 1974, over the signatures of Berg, David Baltimore, and nine other prominent scientists, a single-page letter—sponsored by the U.S. National Academy of Sciences—appeared in the respected English journal *Nature,* and in the American magazine *Science.* It also appeared in the less-widely-read *Proceedings of the National Academy of Sciences.* This unprecedented document called for a worldwide moratorium on a number of intriguing, interesting experiments and, most of all, it sent a tremor—a distinct shock— throughout the scientific community. It was a shock, indeed, even greater than the jolt felt in the early 1940s when physicists reluctantly agreed not to publish their data on nuclear physics, hoping to limit German access to the embryonic lore of atomic weaponry. Back in those war years, however, there was not a call to halt the work; just a plea not to publish.

The biologists' letter warning of the possible hazards of recombining strange bits of DNA was titled simply *Potential Biohazards of Recombinant DNA Molecules*. It asked that proposed experiments be halted if they might introduce genetic resistance to antibiotic drugs into bacteria that weren't yet immune, or give some tame bacterium or virus the ability to cause cancer, or introduce any other "pathogenic" talents into organisms not already possessing such abilities naturally.

An interesting point, in addition to the fact of the moratorium itself, is that during the next two years very few scientists—if any —have violated this voluntary ban on research. Some experimenters had already begun a few of these "forbidden" experiments and were obliged to shelve their plans, ideas, and materials.

Such a moratorium is a rather flimsy, undisciplined structure unless the promise exists that it will end in reasonable time, or that specific rules or guidelines will, in good time, be set up under which at least some of the research can be carried forward.

What this moratorium led to, however, was the renowned conference on recombinant DNA held at Asilomar, California. There, in late February 1975, some 140 of the world's leading bioscientists sat down to argue out the possibilities, the consequences and the probable hazards inherent in continued rejuggling of Mother Nature's living chemical codebook. This conference was funded to the tune of $100,000 by the U.S. National Institutes of Health and the National Science Foundation, and attendance included scientists representing biological research centers in the United States, Japan, England, France, Germany, Russia, and several other nations where such work was being done, or was being contemplated.

Unfortunately, in some ways the Asilomar gathering was too much like other scientific conferences, with individual scientists giving proud, detailed reports on their latest research, backed by slides showing charts, graphs, and magnified strands of DNA. During these presentations some of the very latest work in genetic tinkering was discussed; but the long, dry sessions often seemed to be bypassing the main point of the meeting: assessing the possible dangers of genetic manipulation.

During some of the lively evening sessions, however, scientists, staff from the National Academy of Sciences, and a handful of newswriters downed case after case of Coors beer while the real work of the meeting got under way. Special committees met, haggled over philosophy, practicality, and small details, then finally hammered together their reports. Then, after almost a week of long, dull meetings, sessions of intense negotiating, and lots of midnight rewriting, the conferees finally produced a report which, in essence declared that genetic juggling—with a few important exceptions—should proceed like a pair of porcupines making love: very carefully.

Here are the conference's conclusions, as summarized later by the National Academy of Sciences:

• That most of the research work being done toward building new combinations of DNA molecules should proceed, but with appropriate safeguards. Such safeguards are mainly complex physical barriers designed to prevent the escape of strange—perhaps dangerous—new microorganisms.

• That the standards of protection, both for the workers in the laboratories and the public outside, should be high, especially during the first few years of genetic engineering when very little realistic assessment of the hazards is possible. The feeling was that such standards, even if they are too high at the outset, can be lowered as the assessments of risk change.

• That a few experiments suspected of being ultrahazardous, such as those involving plugging the genes for cancer or a disease like smallpox into normally friendly organisms, should *not* be done under any circumstances at present, and should wait indefinitely until some extra safe new facilities can be built. It was strongly hinted, too, that a few such experiments might be best if never tried at all.

As can be seen, then, the tone of the conference's final declaration was a little on the optimistic side, noting that work in future and more experience would probably show that many of the suspected dangers are actually less serious, less probable than imagined or, even, nonexistent.

To some people, of course, the results of the Asilomar meeting seemed rather self-serving, for the benefit of the biological community. Nonetheless, the conference can still be seen as a significant event in the history of science, an important attempt at self-regulation.

It was the scientists themselves who blew the whistle on their own activities, who showed the responsibility and chose to bring up the question of risk in public. It was the community of scientists, too, which brought itself to suggest—perhaps reluctantly—that some sort of control or regulation might be necessary in the end. One can suspect, too, that one of the things these biologists hoped to do was head off—as directly as possibly—the specter of federal intervention. They weren't interested at all in seeing their research placed under control of a gaggle of bureaucratic federal agencies. Indeed, most of the fears about the rules proposed at Asilomar revolved around the prospects that someday scientists would be forced to wade through endless government red tape just to carry on their research. For scientists, the fear of government control is just as real as it is for American businessmen.

But even if United States agencies become involved in full-scale regulation of the genetic engineering research enterprise, such regulation can't hope to halt—or even slow down—the enthusiastic pace of similar work going on overseas. American guidelines, of course, stop at American borders, and scientists in the United States can only hope that their overseas colleagues take the hint.

Nonetheless, it should be clear that this research *is* going to be done. If not in the United States, then the work will be accomplished in England, Japan, West Germany, or France, and also probably in Russia and Israel. The work is too interesting, too exciting, and too potentially profitable to halt.

But if American agencies do impose too-rigid control over genetic research, actually slowing down some of the most important work, the situation may quickly resemble what has been going on in the pharmaceutical industry, where everything daring and exciting—and important—is being done overseas. Americans are now usually among the last people to benefit from new medicines that might save additional lives. Most specialists blame this lack of initiative on the U.S. Food and Drug Administration (FDA),

which has made certification procedures for new drugs so tedi-
ously slow and expensive.

On the other hand, however, comes the realization that it was
just this kind of red tape, this slowness to approve new drugs,
which saved Americans from the trauma of thousands of mal-
formed babies caused by use of an effective new tranquilizer
called Thalidomide.

In the face of such possibilities or worse, then, strict control of
genetic research by the federal government might not seem all
that onerous, especially to laymen. Indeed, strict and complete
control of such research isn't yet required in all instances, or even
in prospect, but during that Asilomar conference a few observers
from the U.S. National Institutes of Health (NIH) were carefully
taking notes. NIH is the source of funds supporting much of today's biomedical research. Thus the NIH was logically the first
body called on to begin considering some sort of control over
genetic manipulation experiments.

Indeed, NIH committees went through some painful deliberations after the Asilomar meeting and finally did come up with a
set of guidelines for genetic research.

These guidelines cover only those researchers working under
the blessing of NIH grants, who will be obliged to agree to obey
the set of rules issued in mid-1976 by a special NIH advisory committee. Within a few months, too, the same set of guidelines was
adopted by the National Science Foundation, which requires all
its grantees to fall into line, too. In addition, it is hoped that research agencies and their control bodies in other nations will
adopt rules similar to those released by the NIH. In England, for
example, a set of rules was also put into force.

The NIH advisory committee—set up quickly once the potential
dangers of genetic tinkering became well publicized—took pains
to point out early that effective biological safety programs had
been worked out years ago which have successfully protected
most scientists, laboratory workers and the public from serious
infection while at the same time allowing important research to
continue. As a result of earlier precautions—including those involved in handling the agents of biological warfare—some infor-

mation and experience were already available which showed rather clearly how the new work with unfamiliar microorganisms might be done.

These safety-oriented research methods usually involved, first, a rigid set of standard rules of laboratory practices which are already rather well observed in microbiology laboratories. Such roles are generally meant to prevent or at least minimize the chances that something alive and dangerous can escape from the laboratory and be dispersed in the environment. After such rules, the second line of defense involves numerous special procedures, plus specially designed structures and equipment to serve as physical barriers. Such barriers are already present in some laboratories, and in genetic research they would be required in varying degrees according to how dangerous the experiments to be done are thought to be.

Fortunately, too, because of the new techniques involved in genetic engineering, a new type of containment mechanism has become available. This involves the use of biological barriers, which are actually the living bacteria and viruses themselves. Although they are alive and functioning for the purposes of experiment, they are deliberately weakened in advance with specific mutations so they can't survive outside of the special conditions found in the laboratory.

Noting that these three types of safeguards can be complementary, the NIH committee found that they can be combined in different ways for different experiments—designed to meet different levels of potential hazard. In addition, however, the NIH committee emphasized: "The first principle of containment is a strict adherence to good microbiological practices. Consequently, all personnel directly or indirectly [involved] must receive adequate instruction."

Critics of the proposed genetic shuffling experiments answer, however, that this is the area, this "containment" based on the training of human workers, where the new guidelines are apt to first break down. A group of biologists at Harvard University and the Massachusetts Institute of Technology said during public hearings that ". . . some of the sloppiest people in the world are graduate students . . ." who are frantically pressing, frequently in the early morning hours, to finish complex experiments hoping to meet important academic deadlines. Obviously, even in the strictest laboratories, where the rules are rigidly enforced, there will be times when students, professors, or their technicians will be hurried or tired, or both, and will take shortcuts, slighting the precautions if not ignoring the rules altogether. Thus many scientists are convinced that the disastrous release of hazardous, uncontrollable new organisms will come as a result of human error, through some sloppy act, lack of attention, or forgetfulness, rather than through the failure of a physical barrier or the deliberate release of lethal germs. Unfortunately there seems little that can be done, even in the long term, toward avoiding lapses in human judgment, the slip of a hand that can lead to disaster.

Nonetheless, in a careful outline of what these various containment strategies are—strategies which were designed both to minimize danger but still allow important work to continue—the NIH committee divided the physical containment categories into P1, P2, P3 and P4 levels, which can best be defined as standards for containment for experiments believed to have minimal, low, moderate and high risk, respectively.

The P1 or minimal-risk level of biological containment assumes that only relatively safe work will be done, and that the laboratory used is suitably equipped for research involving microorganisms. Still, the laboratory is not required to have any special design features that are required for the more hazardous work. These are requirements for P1 containment:

- Laboratory doors should be kept closed during experiments.
- Work surfaces such as sinks and benches should be decontaminated daily, and immediately after spills of genetic (recombined) materials.
- Liquids containing recombined sets of genes should be fully decontaminated—essentially destroyed—before being dumped down the drain.
- Solid wastes containing new DNA combinations should be decontaminated or sealed inside leakproof containers before being taken out of the laboratory.

• Pipetting (using suction to draw up a sample of fluid inside a glass straw) can be done by mouth. Preferably, however, other means would be used.

• Eating, drinking, smoking, and storage of food in the laboratory working area should be discouraged.

• Handwashing facilities should be available.

• Pests such as rats, mice, and insects should be controlled.

• Laboratory gowns, coats, or uniforms may be used if desired, but are not mandatory.

With such requirements, it should be obvious why P1 containment is considered minimal. The precautions to be followed are basically those followed almost routinely in any modern laboratory where some degree of cleanliness and order are required. That recommendation for a pest-control program, however, raises an interesting point. According to some biologists, a few of the older biological laboratories—including the old biology building at Harvard—have their own unique pest-control problems already. One biologist who formerly did his own research at Harvard said, "There has always been the problem of what we called the radioactive red ants. I found that you can't leave anything edible out on a counter overnight in that building. If you do, there's this long line of tiny red ants stretched out across the counter when you come back. I say they're radioactive because a lot of the solutions used in experiments contain radioactive tracer materials, and the ants must have found their way into that stuff, too."

He added that this one species—or tribe—of red ants seems to be peculiar to biology laboratories.

But ants aside, the next step up the NIH's containment ladder is P2 containment, which the committee described as the type of laboratory "commonly used for experiments involving microorganisms of low biological hazard." Thus the rules for the P2 laboratory, in addition to the requirements listed for P1 containment, require that:

• Only persons who have been warned of the potential biohazards will be allowed in the laboratory.

• Children under age twelve cannot enter the laboratory.

• Solid wastes must be decontaminated or be placed in leakproof containers before removal from the laboratory. Packaged materials must be incinerated or sterilized before disposal by other methods.

• Contaminated equipment and materials—such as glassware which will be reused—must be decontaminated before removal from the laboratory.

• Persons handling recombinant DNA materials should be asked to wash their hands often, especially when they leave the laboratory.

• Laboratory gowns, coats or uniforms are required. These cannot be worn in lunchrooms or outside the laboratory building.

• Animals not needed for experiments should not be allowed in the laboratory.

• Biological safety cabinets and other physical barriers must be used to minimize the hazard of aerosolization (suspension in air) of recombinant DNA molecules.

• Use of hypodermic needles and syringes should be avoided if possible.

Thus, P2 containment is a slightly more expensive proposition than P1 containment, especially since those biological safety cabinets are recommended. And, as should also be expected, the next step up, P3 containment, is even more stringent and more expensive. Under the rules—in addition to the requirements for P1 and P2 laboratories—a P3 laboratory will need numerous features that are specially designed and engineered. Just for openers, a P3 facility must be rigidly, carefully screened to eliminate public access. Such isolation is achieved by controlling entry corridors, by installation of air locks, and with double doors on locker rooms and other facilities. Indeed, all access to the laboratory is under control, and those biological safety cabinets must be available inside the controlled area. An autoclave, which is essentially a pressure cooker designed for sterilizing contaminated equipment, should also be available, preferably within the laboratory itself. Surfaces of walls, bench tops, floors and ceilings must be easily and thoroughly cleanable.

Also required for a P3 laboratory is a system that provides directional air flow within the laboratory. This sophisticated ventilation system must be balanced to draw air in the access corridor into the laboratory. Exhaust air is discharged outside the laboratory and dispersed so it can't reenter the building. No recirculation of the exhaust air will be permitted unless it is properly treated first.

As for the experiments themselves, the NIH's rules for P3 laboratories specify that none of this work will be done using open vessels that contain living organisms carrying new combinations of genes. Nor will this work be done on open laboratory benches. All such procedures are to be conducted inside biological safety cabinets.

In addition, those P3 guidelines require:

• All laboratory access doors will bear the universal biohazard sign. Only persons required for doing the work will enter the laboratory, and they will be warned of the possible hazards before entering. They must also follow all required entry and exit procedures.

• Work surfaces will be decontaminated at the end of each experiment.

• Pipetting by mouth is prohibited.

• Gloves must be worn when handling recombinant DNA materials.

• Laboratory vacuum lines must be fitted with filters and liquid traps.

Before they are removed from biological safety cabinets, all materials must be sterilized or be transferred to unbreakable, sealed containers. These can then be removed from the cabinets through either a decontamination tank, an autoclave, an ultraviolet airlock, or only after the entire cabinet has been decontaminated.

All this makes P3 containment difficult and very expensive, but P4 rules are even more stringent, and extremely restrictive. In most cases, scientists hoping to do experiments needing P4 containment will probably have to wait until new facilities are built, since very few such laboratories exist now, even as the pace of recombinant DNA begins accelerating.

Last in line, then, is P4 containment, and the NIH committee which put the guidelines together warned that experiments requiring P4 containment should only be done in laboratories equipped to handle ". . . microorganisms that are extremely hazardous to man or may cause epidemic disease."

For P4 containment the laboratory—or the entire laboratory building—either is completely separate or is inside a controlled area within a building, and is completely isolated from all other areas of the building. All access to the laboratory is under rigid control.

"A P4 facility," the committee's guidelines said, "has engineering features designed to prevent the escape of microorganisms to the environment. . . ." These include:

• Monolithic walls, ceilings, and floors in which all penetrations for air ducts, electrical cables, and utility pipes are sealed to assure physical isolation.

• Airlocks through which supplies can come safely into the laboratory.

• Double-door autoclaves to sterilize and safely remove wastes and other materials from the laboratory.

• A biowaste treatment system to sterilize liquid effluents.

• A separate ventilation system to maintain negative air pressure and directional air flow.

• A treatment system to decontaminate the air.

• Clothing-change and shower rooms which personnel must use on entering and exiting the building. All workers will shower on exit from the building.

• Complete laboratory clothing, including underwear, pants and shirts or jump suits, shoes, head covers, and gloves must be provided for—and must be used by—all persons entering the laboratory.

With such a list of requirements, it's easy to see why biologists are feeling burdened, but there's even more. As strict and rigid as the P4 requirements may seem, the NIH committee also decided that even P4 containment isn't safe enough for some of the experiments scientists might like to perform. Even while the hazards of these proposed experiments aren't even defined yet, some specialists fear that the escape or spill of a batch of strange new creatures could produce disastrous, even unbelievable results. In other words, the hazards of such forbidden experiments—even if they're still only hypothetical—seem to far outweigh any possible benefits.

Thus far the NIH's list of experiments "Not to Be Done" is still quite small, but the substance of the list, the experiments that are prohibited, provide a strong feeling for what the problems are and where the dangers lie. These prohibitions include:

• No cloning of oncogenic (cancer-causing) DNAs from pathogenic organisms, disease-causing agents which are considered especially dangerous.

• No cloning of oncogenic (cancer-causing) viruses classified by the U.S. National Cancer Institute as even of moderate risk, or of any cells known to be infected by these viruses.

• No deliberate formation of recombined DNAs containing the genes for potent toxins such as those responsible for botulism or diphtheria, or the venoms of snakes and insects.

• No deliberate creation of plant diseases likely to be more damaging or to spread more widely than present plant diseases.

• No deliberate release into the environment of any organism containing a recombinant DNA molecule.

• No transfer of drug resistance traits into microorganisms not now known to acquire them naturally, especially if such resistance might hinder the use of drugs to control diseases in humans, animals or plants.

In addition, the NIH committee decided to prohibit experiments in which large amounts of DNA are used, especially if that DNA is known to code for potentially harmful products. The committee explained its reasoning by noting, "We differentiate between large and small scale experiments because the probabilities of escape from containment barriers normally increases with increasing scale."

A few possibly risky experiments thought to be of direct social value—without a good definition of what "direct social value" might be—could be excepted from this last rule. Large scale experiments were defined as those involving more than ten liters of recombinant DNA. Experiments using greater amounts of material might still be performed, but only with approval from the Re-combinant DNA Molecule Program Advisory Committee of the NIH.

As mentioned before, the NIH took notice that physical con-tainment isn't the only line of defense that can be erected to allow this new gene-shuffling technology to be carried on. Backing up the P1–2–3–4 containment system are newer approaches that scientists are calling biological containment. What this means essentially is that bacteria like *E. coli* might still be used in genetic tinkering experiments, but that some special weakened varieties of these organisms have become available which have minimal chance of surviving outside the laboratory. One of the favorite beasts already in wide use in genetic reconstruction experiments is a variety of *E. coli* known as K–12. This organism has been isolated so long now that the laboratory appears to have become its natural habitat. Indeed, it has been isolated for several decades, and enough mutations have occurred to make it unable to colonize in *E. coli's* favorite haunts, including the human gut. Thus *E. coli* K–12 has already become known as the first, or lowest, level of biological containment, designated EK–1.

The NIH committee noted, however, that its faith in the harm-less character of *E. coli* K–12 extends only so far as the "normal intestine" in humans. Committee members warned that persons being dosed with antibiotic drugs "must not work with DNA re-combinants formed with any *E. coli* K–12 host-vector system during the therapy period and for seven days thereafter."

Similarly, people who have undergone surgical removal of part of the stomach or intestine should avoid work in laboratories using *E. coli,* "as should those who have had large doses of antibiotics."

From EK–1, of course, the next step up the biological contain-ment ladder is EK–2, a classification calling for the use of living organisms which have been purposely disabled genetically. They are meant to self-destruct if they somehow escape and find them-selves loose from the laboratory. The EK–2 requirements order that no more than one bacterial cell out of 100 million should be able to survive—even with some strange bit of new DNA inside—if the bacteria find themselves out free in the natural environment.

Neither should these modified organisms be able to pass bits of new genetic information off to other more normal, uncrippled inhabitants of the natural world.

This "disabling" of *E. coli* and other organisms is accomplished by causing mutations—or by adding mutant DNA—to a normal *E. coli* or to a K–12 type, making it, as one example, unable to survive at normal human body temperatures. Or, alternatively, *E. coli* germs could be made deficient in some important amino acid, or deficient in some critical enzyme. This missing ingredient could be supplied routinely in the laboratory to keep the bacteria healthy, but it would not be abundantly available in a natural setting and they would soon perish. Still another avenue of approach is to make the organism so weak that it can't compete for nutrients with the more vigorous normal residents of the intestine.

As it has turned out, however, such tinkering with the internal life-style of a living organism was more difficult than it seemed at first glance. Toward the end of that famous Asilomar Conference in California the scientists involved were assuming that production of such crippled organisms might take only a few weeks; that they might be ready for testing almost immediately. Unfortunately, this task turned out to be more complex and difficult than expected, so it wasn't until a year later that Dr. Roy Curtiss III, at the University of Alabama, announced he had been able to tailor a new, weakened line of bacteria for use in genetic recombination experiments. Curtiss reported constructing a creature which cannot—at least in theory—survive in the human intestine, cannot tolerate ultraviolet light, is more sensitive than usual to antibiotic drugs, and which reproduces so slowly that it can't begin to keep up with normal bacteria.

When used in combination with the physical barriers being erected in the special laboratories, Curtiss said, the new bacteria "should reduce the probabilities of danger to [other] organisms in the biosphere to an astronomically small value." He added that "I must admit the task has been far more difficult than I and others ever imagined."

Other scientists in other laboratories are also trying to design safe new types of *E. coli* for genetic experiments. The NIH urged:

"When any investigator has obtained data on the level of containment produced by a proposed EK–2 system, these should be reported as rapidly as possible to permit general awareness and evaluation of the safety features of the new system. Investigators are also encouraged to make such new safer cloning systems generally available to other scientists."

Also in these NIH guidelines is mention of an even higher level of biological containment, EK–3. This category would be basically the same as EK–2 except that the weaknesses designed into the living organism would have been confirmed independently of the original laboratory tests. Independent confirmation must include trials for survivability in the intestinal tracts of higher animals, including the monkeys and apes, and in what are thought to be other "relevant environments." For the time being, the committee said, no weakened bacterium or virus will be considered for EK–3 status without certification by the NIH's Recombinant DNA Molecule Program Advisory Committee.

As should be expected, issuance of this complex set of guidelines in the early summer of 1976 met with most of the possible, predictable responses. Critics who were originally loudly opposed to the whole question remained critical. Those who were largely neutral toward the craft of genetic tinkering saw the rules as an important step, an appropriate step, in the right direction. And those who wanted no rules at all promptly found the new guidelines much too tough, much too restrictive.

American drug and chemical companies generally supported the NIH committee's ideas about containment and other obvious safety measures, but these commercial interests also began fretting seriously over what they considered rules almost tailor-made to suit the needs of the academic community. Some grumbling was heard about these guidelines not meshing well with the needs of the larger, more mission-oriented industrial laboratories. Some twenty representatives of America's powerful pharmaceutical industry soon met with the NIH's director, Donald Fredrickson—even before the NIH guidelines were publicly released—to voice their objections, feelings, and fears.

Speaking for the Pharmaceutical Manufacturers Association,

Dr. John G. Adams hinted that his organization might sponsor its own special meeting which would be aimed at setting up industry's own separate set of guidelines which would be more in harmony with commercial requirements. Indeed, by the time the NIH guidelines were ready for release, at least six large firms had already embarked on active genetic research programs using the newest techniques, and they were essentially free of regulation since the guidelines cover only work funded by the NIH. Companies engaging in this work included such industrial giants as Eli Lilly, Merck, Upjohn, Hoffman-La Roche, Smith, Kline and French, and Miles Laboratories.

The guidelines are extendable, of course, to cover areas beyond the NIH's domain. The U.S. National Science Foundation, for instance, decided early that its own projects, those supported by NSF money, would also be governed by the new guidelines.

Industry's main objections were focussed basically on three points:

• That even though these new rules bear the title "guidelines," the sneaking suspicion is that they could easily acquire the weight of fully enforceable regulations. The businessmen were worried that the provisions enforced might not be in the best interests of the drug industry.

• That the new set of guidelines, calling for a voluntary national registry of genetic manipulation experiments, are likely to run afoul of industry's obvious need for secrecy during development of important new products.

• That some of the new NIH guidelines might cause hardship or delays for industrial researchers because of the restrictions forbidding large-scale experiments, those which involve large amounts of recombined—rebuilt—genetic material. For industry, certainly, large amounts of biological material are frequently needed for assessing the commercial feasibility of any given project.

Such criticism from all sides was almost constant during the whole time the NIH committee was working to formulate guidelines. As first proposed—several months after that important meeting at Asilomar—the gestating set of rules was quickly, roundly denounced as being too lenient, not at all meeting the suggested limits that emerged, however imperfectly, from the Asilomar experience. Some biologists cautioned that the world would look at a set of lenient rules as designed to serve only the scientific community. The NIH committee thus went back to the drawing board and began building the hazy, generalized rules spawned at Asilomar into a more concise, clear set of guidelines that could be backed up by the strong threat of funding restrictions.

Meanwhile, in England, the going wasn't much smoother. Just prior to that Asilomar meeting the first public report on the new recombinant DNA work was issued by a committee under the leadership of Lord Ashby, botanist, master of Cambridge University's Clare College. The Ashby report was thus available at the time of the big Asilomar meeting—in February 1975—to provide an important guide for the generally meandering discussions in California.

The Ashby committee compiled its report after consideration of evidence submitted by twenty-eight British experts in molecular biology. In part, the Ashby report declared:

> We were not set up to prepare a code of practice for workers who use these [genetic engineering] techniques. Nor were we set up to make ethical judgments about the use of techniques. Our business has been to assess the potential benefits and hazards after discussion with scientists who are familiar with this branch of biology.

Nonetheless, the Ashby committee did suggest two ways in which the hazards of research using novel, untried new kinds of DNA might be conducted in a safer manner. First, the committee wrote, "is to accept the hazard and contain it." The other is to reduce or eliminate the hazard by, as one witness put it, "disarming the bug."

Before the end of 1976, however, the first hints that some form of international regulation of genetic tinkering might be possible began to emerge from the respected International Council of Scientific Unions (ICSU). At its sixteenth General Assembly, held in Washington, D.C., in October, the council set up a special new committee that was asked to censure any nations which refuse to observe the safety standards prevalent in the majority of other nations doing recombinant DNA work. The new organization is called the Committee on Genetic Experimentation, and its duties include watching how all governments act in regard to genetic research, encouraging governments to set up uniform research policies, and providing the public, worldwide, with reports on the benefits, hazards, legal problems and ethical issues involved in recombinant DNA experiments.

Unfortunately, this new ICSU committee will have no legal enforcement powers—no teeth—and Dr. Philip Handler, who headed the United States delegation, commented, "The only force we have is moral persuasion."

W. J. Whelan, the chairman of the ICSU's ad hoc committee on recombinant DNA molecules, also pointed out that the committee does not believe genetic engineering experiments should be halted, but that "the committee is interested in seeing research go ahead, but responsibly."

The Ashby committee, however, had first suggested what was to become one of the key goals announced after the Asilomar conference: the creation of that significantly sickly bug.

Even after creation of such weakened bacteria, when scientists thought they had finally found a good vehicle for continued recombinant DNA research, a large squad of fellow scientists still remained unconvinced, still contending—along with Chargaff— that *any* strain of *E. coli* is too dangerous to use in these experiments. According to Dr. Stanley Falkow of the University of Washington, an almost astronomical number of different *E. coli* strains exists, some of which differ by as much as 25 percent in genetic code sequences. There are some *E. coli* strains, too, which are known to cause infantile diarrhea and other diseases resembling dysentery—the famous traveler's diarrhea or "Montezuma's revenge"—that so many Westerners suffer during visits to the developing countries. It is also known that *E. coli*, while not normally considered a highly virulent organism, does occasionally acquire additional genetic talent—usually through transfer of a new plasmid—which can tip the balance between being an irrelevant, benign inhabitant of the human intestine toward having the ability to touch off a case of serious disease.

Scientists involved in experiments using *E. coli* as the genetic workhorse retort, however, that their favorite tool has been a laboratory captive so long that it is no longer able to survive in the wild. Tests done on human volunteers appear to prove this contention correct. Nonetheless, Falkow points out that other experience with *E. coli* K–12 hints that it *can*, under the right circumstances, be strengthened, be made more viable, through transfer of genes from another bacterium via a plasmid. A dose of just the right additional genetic information, Falkow said, "could probably affect the ability of this *E. coli* strain to survive in the gastrointestinal tract."

E. coli has also been described as exhibiting "untiring promiscuity" as it exchanges bits of genetic material with a number of other different organisms, sometimes sending its own plasmids off to grow in other types of bacteria.

Defenders of this research are quick to reply that such concerns over safety, however well-intended, have little if any realistic basis in fact. They argue that the few hypothetical cases discussed by critics would require the precise, delicate coordination of many improbable events, meaning that the overall chances of such occurrences, even in *E. coli*, are truly negligible.

Some of the scientists who are building their careers on this genetic recombination work also claim that the use of genes taken from higher animals, inserted into bacteria, is probably also less dangerous than some people imagine, simply because such genes are very rarely, if ever, expressed or acted upon by the bacterium's internal machinery. In order to be expressed, the higher organism's genes require that complex control sequences, or codes, be read by the bacterium's genetic tape recorder, and so far the bacterium doesn't appear to be able to do this.

In addition, a few scientists have forcefully pointed out that there is normally so much DNA floating around in the world's waters that almost all—if not all—possible combinations of genes have at one time or another already been tried by Mother Nature, and most such combinations have been found useless.

Given such arguments, then, the dangers of genetic tinkering are difficult to assess when it comes to putting together a reasonable set of research guidelines. Fredrickson, director of NIH, also noted in a lengthy statement released with the new guidelines that the whole science of genetic tinkering—and its potential hazards—are extremely difficult to communicate to the public. He added, too, that it is equally difficult to arrange for public participation in the decision-making process.

He explained why by noting, "The field of research involved is a complicated one, at the leading edge of biological science. The experiments are extremely technical and complex. Molecular biologists active in this research have means of keeping informed, but even they may fail to keep abreast of the newest developments. It is not surprising that scientists in other fields and the general public have difficulty in understanding advances in recombinant DNA research. Yet public awareness and understanding of this line of investigation is vital."

Despite the obstacles, that NIH committee was formed in October 1974 to advise the Secretary of Health, Education and Welfare, the assistant secretary for health, and the director of NIH on how to develop some sort of procedures—also involving the public—that will minimize or avoid the spread of strange new DNA combinations "within human and other populations" and to come up with guidelines to be followed by scientists doing those possibly dangerous experiments.

Fredrickson took pains to point out that he believes "public responsibility . . . weighs heavily in this genetic research area. The scientific community must have the public's confidence that the goals of this profoundly important research accord respect in important ethical, legal and social values of our society."

He also stressed that it is important for the advisory group's meetings to be characterized by openness and candor, and that all important points of view be represented.

The primary objective of the NIH guidelines, then, Fredrickson said, "is to ensure that experimental DNA recombination will have no ill effects on those engaged in the work, on the general public or on the environment."

One important point observed during the process of putting that set of guidelines together, Fredrickson added, is that "practically all commentators supported the present prohibition of certain experiments" such as the attempts to give some bacteria new types of resistance to antibiotic drugs. Nonetheless, this notion also spawned its own share of argument because some respected, careful scientists had already been using this quality, this talent known as drug resistance, as a convenient "marker" to test whether an attempt at genetic change had been successful. Still, the NIH committee, as mentioned earlier, decided to add prohibitions against such experiments which might give disease organisms better armament against man's best weapons, and the committee also decided to prohibit experiments which might lend bacteria the ability to make some of the deadly venoms found in snakes and insects.

In warning that the new guidelines also prohibit release of newly created or altered organisms into the environment, Fredrickson declared:

> I have decided that the guidelines should, for the present, prohibit any deliberate release of organisms containing recombinant DNA into the environment. With the present limited state of knowledge, it seems unlikely there will be, in the near future, any recombinant organism that is accepted as being beneficial to introduce into the environment.
>
> When the scientific evidence becomes available that the potential benefits of recombinant organisms, particularly for agriculture, are about to be realized, then the guidelines can be altered to meet the needs for release.

Fredrickson also stated—mostly in answer to critics who fear the use of *E. coli* as a recombinant DNA vehicle—that "for years it [*E. coli* K–12] has been the subject of more intense investigation than any other single organism, and knowledge of its genetic makeup and recombinant behavior exceeds greatly that pertaining

to any other organism. I believe that because of this experience
E. coli K–12 will provide a host-vector system that is safer than
other candidate microorganisms."

Fredrickson also declared, however, that additional study
should be pushed so that other organisms such as *Bacillus subtilis*
—which has no comfortable ecological niche in humans—can be
perfected for such experiments. Of course the main problem is
that development of new strains of bugs for use in research isn't
a very glamorous occupation. Asking young scientists to pass up
more exciting work to take on this task won't be easy.

As explained earlier, too, the new NIH guidelines have been
made rather explicit about the different levels of containment
facilities required if foreign DNA is to be stuffed into *E. coli*. And
these requirements, as mentioned, become stiffer—demanding ever
more secure facilities—as the DNA being inserted into bacteria
becomes further and further removed from *E. coli* in terms of
evolution. This is important because it must be assumed that the
sets of genes lifted out of some higher animal might in some unexpected cases be expressed even inside the primitive bacterium,
with the genetic information from the animal being decoded and
obeyed. This assumption, remote as it is, boosts the need for
greater caution.

"It is further assumed," Fredrickson stated, "that the product of
that foreign gene [inserted into *E. coli*] would be most harmful to
man if it were [coding for] an enzyme, hormone, or other protein
that was similar to proteins already produced in man. An example
is a bacterium that could produce insulin. Such a 'rogue' bacterium could be of benefit if contained; a nuisance or possibly
dangerous if capable of surviving in nature."

Another and perhaps even better reason for scaling up security
levels when higher animals' genes are inserted into bacteria is the
concern that tiny viruses hidden among these new genes might be
carried into *E. coli* during such experiments. Once inside the
bacterium, it's conceivable that such hidden viruses—indeed, cancer viruses—could be reproduced in huge numbers, and that they
could later be released to cause disease. Or worse, they could

perhaps give their host bacteria the power to hit humans with a new disease, or a new form of an old disease, or even cancer.

Such risks are considered even more hazardous if the scientist is busily inserting large, rather ill-defined segments of DNA into his bacteria. Such experiments, while considered valuable in some instances, have become known as "shotgun experiments" because it is impossible to forecast what products, what genetic combinations, will come out of such attempts at tinkering. The end results could—and probably will—turn out to be mostly benign, but the chance still exists that something deadly could be made, and the danger might depend on where the particular type of bacterium used as a host likes to make its home. If "home" is the human gut, then problems might be expected.

How this can happen is probably best illustrated by the story of one widely used vaccine. After millions of persons had been inoculated with the medicine, researchers discovered that it was contaminated with a monkey virus known as SV–40 (simian virus–40). The vaccine had been grown on tissue cultures of monkey cells, and the vaccine had apparently picked up the genetic instructions for the virus from the DNA in monkey cells. Fortunately, the monkey virus doesn't appear to be able to grow and cause any damage in humans, and more than twenty years have now passed since that laboratory "accident" occurred. Nonetheless, it is a somewhat frightening example—perhaps even a warning—about what can happen when scientists begin manipulating living things.

As the era of genetic tinkering opened, we should have expected that all the publicity and all the concerns about safety would soon find their way into the public arena where politicians might also become involved. That, indeed, is what has happened, and the first place where political interest in the questions began making news was at the University of Michigan in Ann Arbor. Biologists at work in the university's laboratories—like biologists working elsewhere—were anxious to move forward in their exciting work, and in some cases their plans called for insertion of strange new genes into *E. coli*. As was beginning to happen at other universities, critics there were beginning to voice fears about the wisdom of conducting potentially disastrous experiments in the middle of a large community. The question finally came before the university's regents who agreed, after much discussion, that the research should continue. The regents also recommended, however, that no "high-risk" experiments be done, and that only specifically enfeebled bacteria be used as the recipients for new genes.

One important new element arose during the debate in Ann Arbor, however, in that the true representatives of the public—the city's elected leaders—chose for the first time to listen in on the fight. While not actually involved in the decision—and feeling rather left out and powerless because of it—the city's representatives did demonstrate the first real awakening of local political interest in the hazards of genetic recombination. Thus the often-strained town-gown relationship has found one more strong source of tension.

After the decision to continue genetic tinkering research was made in Michigan, however, the debate moved quickly to Cambridge, Massachusetts, home of Harvard University and the Massachusetts Institute of Technology. There, the questions about safety first arose at Harvard, then spread almost automatically to M.I.T., after the Harvard biology faculty announced plans for building a special containment laboratory for recombinant DNA work on the fourth floor of the university's biology building.

If the laboratory couldn't be built there, the alternative choice was to build new laboratories some distance away in an unused atomic accelerator building. This second choice would mean that researchers doing biology experiments would be traveling away from their own offices and laboratories for their research work, a prospect they didn't like.

Whether they like it or not, however, the critics were still arguing against the use of the biology building, right in the heart of Cambridge, as the place for building new containment laboratories. They soon began raising the specter of strange new diseases erupting in Harvard Square. They also contacted the Cambridge City Council, and found this was one means by which they could drag the growing dispute into an open forum. The first result was to quickly slow down progress toward using those new laboratories.

Nonetheless, in the meantime Harvard University officials, after deliberations by several select committees, decided to go ahead and build new P–3 containment laboratories in the biology building. The feeling in Harvard circles was that they were actually doing more than they had to, since P–3 containment was considerably more restrictive and severe than the experiments they planned to do required. Nonetheless, a firm "whoa" soon came from the city council, and the argument soon developed into an extraordinary confrontation in the Cambridge City Council chambers. This, then, represented the first real public collision between genetic scientists—especially those who want to do the tinkering—and the community in which they work.

Perhaps it was just by coincidence, but this clash in the Cambridge council chambers occurred on the very day—in the late spring of 1976—when Fredrickson, at the National Institutes of Health in Washington, was handing out his agency's new set of genetic research guidelines. Thus the contentious hearings in Cambridge were already providing a first test of public acceptability of the federal rules. And from all indications, acceptance was neither immediate nor unanimous. In the end, the council voted to ask Harvard and M.I.T. to observe a six-month "good faith" moratorium on recombinant DNA experimentation which might require either P–3 or P–4 levels of containment. The council then asked the city manager, James Leo Sullivan, to set up a special Cambridge Laboratory Experimentation Review Board to explore the whole question all over again. It was also suggested that in the future the city keep in close contact with the special safety committees at both Harvard and M.I.T. to help oversee what genetic research is done, and under what conditions.

Finally, in February 1977, the Cambridge politicians agreed unanimously that the work can continue at Harvard and M.I.T. In the process of agreeing, however, council members extended the NIH guidelines to everyone doing the work in Cambridge, including private industrial laboratories. An important point, too, is that the Cambridge council was the first such body to put the weight of law behind those guidelines.

As a result, it would seem that one of the things the scientists fear most is occurring. It may soon turn out that complex biological research is being overseen by outside agencies which naturally have more concern for political expediency than for scientific purity.

One consequence of the Cambridge City Council's actions, according to Graham Chedd, writing for *New Scientist* magazine, is that the precedent established in Cambridge will probably extend much beyond Cambridge, probably to most of the cities which play host to major American universities.

One of the important reasons this intense argument stood out so vividly in Cambridge, however, is because Harvard University serves as headquarters for an important, vocal and articulate organization of ultraliberal-leaning researchers known as Science for the People, a group which is passionately opposed to present approaches to the control of—or lack of control of—genetic manipulation. Indeed, members of Science for the People stress the point that scientists, most of all, should *not* be left by themselves to make the important decisions on what types of genetic tinkering should be done and how they should be done. Members of this advocacy group claim that scientists, since they are so tied up in the lore of the laboratory, are unable to see much beyond their own research, and their decisions would be colored by their desire to get on with the work.

Jonathan King, a member of the faculty at M.I.T., spoke for Science for the People when he argued that the debate in the Cambridge City Council chambers was the ideal forum in which the squabble could be thrashed out—in public—since the issues have become "political rather than scientific." King also stated bluntly that the last person he would trust to protect his health "is someone who doesn't believe there *is* a threat to my health." In a democracy, he added, only the community should have the right to decide whether or not a perhaps hazardous type of work is worth its supposed benefits.

Another opponent of unfettered genetic recombinant research being done in the heart of town was Dr. Ruth Hubbard, who occupies rooms on the fourth floor of Harvard's biology building and who is also associated with the Marine Biological Laboratory at Woods Hole, Massachusetts. Her main argument, also voiced before the Cambridge City Council as well as at the meetings of other committees, is that *E. coli* should not be the host organism for these experiments. She is convinced that living *E. coli* bacteria will eventually be carried out of a P–3 laboratory on someone's skin or clothing, and that "*E. coli* can be all over the place before we know it is out—and we cannot call it back."

In a letter to the editor of the *Boston Globe*, Dr. Hubbard responded to an earlier letter from M.I.T.'s Dr. Salvadore Luria on the threats to intellectual freedom, declaring:

> I would like to take issue with Prof. Luria's plea for the intellectual freedom to pursue research on recombinant DNA.
>
> I am opposed to this work or to having it done in crowded cities like Cambridge because I believe it is potentially dangerous. No one knows exactly how great the danger is and obviously the people who want to do the work say the risks are minimal; but no one denies the existence of a potential hazard.
>
> Given this fact, the path of prudence is to restrict the work to one or a few laboratories away from centers of population, to which access is restricted to only the people who work there, to monitor the health of those people and their families, and to spend the next few years doing the kinds of experiments that will begin to tell us what the hazards are and how to control them.
>
> If the research proves as important as its proponents claim, it will still be important in a few years; but during that time one could learn enough to make it a great deal safer.
>
> If, on the other hand, carefully monitored experiments reveal a considerable level of hazard, this line of work probably should not continue at all and should surely not be permitted to be done in large numbers of university and industrial laboratories across the country.
>
> What is at stake is not freedom of inquiry, but the freedom of a few scientists in a hurry, to endanger other people.

Dr. Hubbard also made statements about the potential for the spread of new recombinant versions of *E. coli*, statements which are perhaps the most telling part of her argument. Not only does

E. coli normally colonize the human intestine, it is also found in nasal passages in humans, in the intestines of other warm-blooded animals, in fish, in insects such as beetles, grasshoppers and flies, and *E. coli* is also found alive and viable in the soil in most regions of the earth, whether sparsely or densely populated.

Now, in view of the Cambridge experience and the spread of concern to other university centers, it seems true that the controversy over genetic manipulation will probably also be spreading abroad to Europe, Japan and other alert, research-conscious nations where public dissent is allowed. Indeed, Charles Weissmann, head of the European Molecular Biology Organization's committee on genetic engineering has already lamented that public reaction in Europe seems to be getting out of hand, that people there are now beginning to see the dangers out of proportion with the risks they normally face in daily life.

This, of course, is the same basic argument the nuclear industry is using in combat against critics who claim that atomic energy is too risky. Thus the two battles—the fight over genetic engineering and the fears about atomic power—appear to be following roughly parallel courses, with an awakening public being tossed between overemotional claims of danger on one hand and the overconfident reassurances from "experts" on the other.

One thing however seems clear. As suggested by Caltech's Dr. Robert Sinsheimer, the wisest course for today's genetic tinkering does appear to be to isolate this work, to send it off to some ultra-secure laboratory—Sinsheimer suggests the Army's old biowarfare center, Fort Detrick, Maryland—until the questions are better answered. The analogous problem, atomic energy, is already operating under this type of restraint. Research was eventually moved to Oak Ridge in the backwoods of Tennessee and to Hanford in the empty western reaches of Washington State. Even now, atomic licensing boards will not allow commercial power reactors to be built and operated in the centers of large population areas.

Obviously, since the accidental release of a lethal new strain of disease could, in theory, be as disastrous as the major failure of a nuclear power plant, the same sort of precautions would seem to be in order.

Too, on the question of whether genetic engineering is too dangerous to do at all, the true answer probably lies somewhere between the two extremes of "don't" and "hurry up." It seems probable that for both—genetics and atomics—the issues are going to be debated in the village square. The people, one way or another, are going to be heard. One hopes it's not too late.

9

Ethics and Issues: Should We Sidestep Evolution?

IT'S BEEN a long, difficult road, a long journey up through the mud and mire of evolution, up through 750,000 generations of mostly ordinary people, of too many villains and not enough saints, since the "birth" of man. But we made it, and today we stand, here on the palace threshold, arrogantly planning a new and subtle biological coup d'etat which, once and for all, may topple Mother Nature from her throne.

Do we know what we're doing?

Probably not. Nonetheless, it's happening. The cannons have been rolled out, the bombs are in place, and the first fuse has now been lit.

So there it sits, sputtering, dropping sparks as a pretty, elegant touch of fire creeps slowly, inexorably up the fuse toward a bigger, stranger "bang" than anyone's ever imagined.

Temporarily, at least, this new, exciting revolution will proceed under the wholesome banner of genetic engineering, but what it really represents is another of this human species' overconfident attempts at playing God.

It may even work better this time, but don't hold your breath. Mother Nature still has an array of weapons, devices, strategies, and even some armies we haven't dreamed of yet. There are going to be some surprises.

But that doesn't seem to matter. Genetic engineering—the shuffling, mixing and matching of unrelated genes, life's most basic set of blueprints—has arrived, like it or not. We can't go back. The research has been done, the results have been published. It's more a matter of technology now than of science. Only a few more important steps, and a number of lucky breaks, now stand between man, a product of evolution, and man, the would-be Creator.

So, given the reality of genetic engineering, what are we going to do with it? First, the important questions:

1. Who's going to make the decisions? Who, as we take command of the genes, gets to play God? Who decides what's normal, what's abnormal, what's good, what's bad?

2. Does man have the right—much less the duty—to barge in on a delicate worldwide evolutionary system that has been running along nicely—even if somewhat inefficiently—for 4 billion years?

3. Can we actually expect to do Mother Nature's job better than she can? Or does nature, indeed, know best?

4. When we speak of control—perhaps legal control—over genetic engineering, what do we mean? What kind of control, if any, is possible?

5. What, if any, are the goals of genetic engineering? Who sets these goals? Are they attainable?

6. What about freedom? Will genetic engineering grant more freedom to more people? Or do we risk enslaving everyone?

7. When those decisions about what's done with genetic engineering are made, will they be made for the right reasons?

8. Who benefits? Who gets hurt?

9. Is genetic engineering worth the risk? Do the benefits outweigh the costs, or vice versa?

10. What should we do?

First, and perhaps foremost, it should be clear there are no answers to any of these questions. There are opinions—piles and piles of opinions—each as good as the next, each as bad as the other. But there are no answers.

Now, the decisions. It's obvious for now, and probably for the next few years, that nobody—including the scientists doing the work—knows what the important decisions in genetic engineering are going to be. One decision, however, a decision that's already been made—to begin this science with a voluntary moratorium on some types of dangerous experiments—was a momentous and valuable example of responsibility which should be applauded worldwide. Few persons, including scientists, have ever been able to so discipline themselves, to say, "Stop, let's look at what we're doing," while poised excitedly right on the brink of a grand new adventure. But they did it, and it worked, and the direct result is a new set of guidelines which *do* set up a fragile framework for making some of the choices. Those "thou shalt not" experiments listed in the guidelines are one case in point.

A flock of small decisions—those being made by individual scientists working in their own laboratories—are for the present going to be most important. No machinery suitable for the task of overseeing all of this genetic recombination work is yet in place, so it remains for individual research workers to decide on their own which genes to plug into which hosts for what reasons. Usually it will boil down to experimenters doing the experiments that are the easiest to do, which give the clearest results and provide important, impressive new answers. The results of such experiments will then lead to other experiments in search of other answers, and these in turn will also help determine what directions the research follows.

Thus what can be expected is a rather quiet evolution of the information, with bits added here, pieces plugged in there, until a whole fabric of research presents itself, nearly full-blown, to an unready world.

But nobody can successfully attempt to rigidly guide pure, basic research without stifling research in the effort. The scientific quest remains a matter of curiosity, a matter of one scientist or group of scientists in one laboratory, other scientists in a different laboratory, becoming intrigued with difficult, interesting problems and deciding to work them out. The notion that money poured into any of these projects can buy quick, accurate results is a fallacy. Without the hard, basic research being done methodically, well in advance, there is very little that applied research can work with. Indeed, a basic understanding has to come first.

A good example of how this works was the huge, successful American space program of the late 1950s and 1960s. A lot of interesting new science came out of the program, but the whole effort would probably have been futile had not the groundwork been well laid in the 1920s, 1930s, and 1940s by scientists who—because of curiosity—had gone ahead and done the basic research that led to computers, to the chemistry of rocket fuels and to the development of tough new ceramics and metals.

An example of a failure in government-pushed applied research —the pouring of money on a problem—was President Richard Nixon's grand and famous war on cancer. The $100 million allocated was aimed at somehow solving the problem of malignancy, but the money went instead into force-feeding a flock of troops who had no weapons. The truth about cancer is that nobody really knows yet what it is, how it starts, grows, gets out of control and, sometimes, is cured. Until the basic biological and biochemical research is done, cancer probably won't be cured no matter how much hard, cold cash is thrown at it.

So it is, too, with genetic engineering. If government decides to force it in one direction or another, using stiff regulations or even "carrot money," one branch or another withers as others are encouraged, and thus some of the important research doesn't get done while other, perhaps less useful areas are overdone. This has happened before in biology, with some fashionable research areas being pushed while others were neglected. Membrane research, for instance, lay virtually dormant for years as biologists dug excitedly into the mysteries and challenge of the cell's nucleus. Suddenly, just in the past two decades, a new awakening has led researchers to realize that membranes, which serve as the sensitive boundaries both inside and outside the living cell, aren't well known, and that this lack of knowledge has become an obstacle blocking a more complete understanding of the cell. Thus, membranes have now become one of the fashionable, challenging areas of research, a lot of research that is long overdue. The same, now, is also true of the neurosciences, although better understanding of the brain and the rest of the nervous system was blocked for many years by the sheer complexity of the system itself.

Tools, too, help make the decisions in science. New ways for measuring things, for controlling and understanding chemical reactions and for "seeing" what goes on inside living organisms often spur an excited burst of research in new directions, sometimes in directions that nobody really expected.

Now, of course, much of the research in genetic engineering has already been done. So, once the scientists finally hand over the ability for significantly altering the genes of both the microbes and the higher animals—including man—who decides what gets altered, and in what ways?

The answers will be important. According to some scientists, genetic engineering is going to be so easy to do that even relatively primitive laboratories are going to be able to create their own strange new forms of life, for whatever reasons. Does this mean that someone like Uganda's quixotic dictator, Idi Amin Dada, is going to have the tools—perhaps at Makerere University —to design and build strange new diseases that he can scatter among his foreign enemies, regardless of the worldwide consequences? Is there any doubt that Adolf Hitler—infamous for his "final solution" for the existence of Jews—would soon have found some horribly imaginative uses for the technology of genetic engineering? Obviously, Hitler would have seen this bit of science as a godsend for his planned "improvements" of the so-called Aryan race, as his breeding program would suggest. But what else would he have done? Considering such questions is no treat for the imagination.

One can ask, too, how the new tools and techniques of genetic engineering might be used in a modern version of the Inquisition. Dressed in the authoritarian robes of religion, anything might be possible with genetic engineering, including the long-term goal of building up a large, docile, right-thinking population of obedient believers. Not that churchmen today would want that, but again, we're looking at some rather extreme possibilities, especially when dealing with mind control. We're already reaching the point, indeed, where "right-thinking" may soon be possible to impose through psychoactive drugs, administered somehow on a massive scale. It's also conceivable that the tricks of genetic engineering, practiced long and hard, could be used as one method for gaining permanent control of behavior. But nobody in his right mind, as they say, wants that.

With the world divided up as it is now, however, it's not too difficult to look ahead a little and sort out some hints about how the decision making in genetic engineering is going to go.

Western nations like the United States, Britain, Germany, and France will be getting into the genetic engineering business earliest and deepest. In response to the challenges and the obvious dangers, they will dither around, playing ideas against other ideas, hold conferences, have protests and finally reach an unhappy consensus about what to do and what not to do, probably pleasing no one but the bureaucrats. Some of these decisions may even be forced by people taking to the streets, marching in response to events that have starkly pointed up the dangers of uncontrolled genetic manipulation. Important progress toward supplying new food to the hungry, new medicine for the ill, and new hope for the genetically handicapped will be slowed, but the work will continue, barring any unforeseen disasters such as a third world war. Under the flag of free enterprise, Western industry will continue doing what it does best, using whatever in genetic engineering is profitable, ignoring what isn't. Because of this approach, industry will end up making a few rather costly mistakes, but will probably cause no disasters.

The Communist nations, along with a few Third World countries, will also be pushing genetic engineering research, if just in the hope of remaining competitive. Russians, in view of the disastrous Lysenko Affair, will be looking at genetic manipulation as one means of catching up with the West in the agricultural sciences. But in all of these nations the decisions probably won't be made in public. Protests, demonstrations and other acts of disobedience probably won't ever be organized, and if they are, will be suppressed and ignored. Like private enterprise in the West, Communist nations will be selecting whatever they can use—in terms of ideology and industry—from genetic engineering, ignoring the rest.

On both sides of the Iron Curtain the decisions concerning what aspects of genetic engineering are good and what aspects are bad will be made along existing traditional lines, probably by the same kinds of people who are already in power, and with a tendency to maintain the status quo. Thus one shouldn't look for any great ethical breakthroughs. In the United States, for instance, a people unable to deal successfully with the issues of abortion, prayers in the schools, and racial integration aren't suddenly going to neatly resolve the moral and ethical dilemmas tangled up in genetic engineering. It now seems almost certain that Dr. Frankenstein's unlovely little monsters may indeed get built—perhaps by accident—before Americans wake up to the reality of this slumbering new science and begin to come up with some answers. Waiting so long will be unfortunate, but it may not be too late. Still, nobody knows.

An interesting point in this whole topic is that many people fear that some government—or worse, its unbridled intelligence agency —will decide it's a wise and wonderful thing to design and build the "perfect soldier," then try to clone up a million copies of this creature. Indeed, this may become possible, but it's not very probable. As technology is developing now, it seems much more likely that automated machines will be more useful, more dependable, faster and cheaper to use in warfare than any such army of subhuman (or superhuman) zombies. The basic problem is that each "perfect soldier" would still have to eat, would need clothing and would have to have shelter. Machines don't, and they seldom talk back.

More likely is that when the cloning of human beings finally begins—if ever—the individuals cloned will probably be examples of the best and brightest, rather than a flock of dullards suited only for cannon fodder.

So, once given these new abilities to transfer genes back and forth between organisms, probably violating near-sacred barriers established long ago by nature, what will be man's responsibility to the rest of the creatures on earth? Do we have the right to change, even by accident, the basic conditions we inherited with this planet?

Who can say? Answers will depend on one's point of view, on one's heritage and one's philosophy. But what can be said, certainly, is that man, through many of his activities, has been changing the earth for centuries, often in massive ways, other times in quiet, hidden ways that even now go unnoticed.

For example, man is a master builder of deserts. Now vast areas of the Middle East, the very cradle of Western civilization, lie barren and useless, unproductive after centuries of unwise use—poor irrigation, overgrazing, and deforestation have left the soil barren and dry, thick with salts. Bare, rocky hillsides in Lebanon, now nearly devoid of those famous, majestic cedars, give testimony to the efficacy of man's ax-wielding industry. Spain, too, is barren of her formerly extensive forests, thanks to shipbuilders, and Scotland's once-majestic Caledonian Forest has all but disappeared.

Look, too, at earth's atmosphere. Is there any doubt that man's hurried activities have ruined the climate that made southern California famous? And what happens next as man, rapidly burning up the world's stock of fossil fuels, keeps enriching the air with carbon dioxide, blocking the reflection of infrared light, warming up the atmosphere? Do man's stirred-up dust particles and ashes compensate, causing the opposite effect by blocking sunlight from reaching the surface? What happens to ice caps at the poles, to sea levels, and to coastal cities if we change the temperature of the earth too much in either direction?

And then look at the impacts of some other human activities. For the mere sake of convenience, millions of Americans—and a lot of Europeans, too—now joyfully spray their hair with goo, and squirt their bodies with strange deodorant chemicals, hoping to render themselves more "acceptable" to others. But is it acceptable to use these very stable gases in spray cans for such trivial reasons when it is now strongly suspected—and virtually documented—that freons destroy the layer of ozone gas high in the atmosphere which normally protects all of us from ultraviolet light? The result, given time enough and bureaucratic lethargy, will probably be increased numbers of skin cancer cases, especially among caucasians. Is that, then, a reasonable price to pay for the convenience of spray cans?

All these, of course, are possible man-made changes of massive proportions on the earth. Yet we seldom sit back and carefully assess the dangers, the destruction, wrought by man's carelessness. What are the chances, then, that we will start now, with genetic engineering, looking at the consequences of our actions? By his very nature, man seems destined to drastically change the earth. We've made animals extinct; we've eliminated swamps, forests, and beaches. We're now poisoning the seas, the land, and the air.

Thus it's perhaps not fair to be asking the bioscientists if they've fully considered what their interesting work might do to man's fragile habitat. Maybe it's up to these scientists to remind the rest of mankind strongly about the human animal's prior and continuing abuses and thus help put things into better perspective. Even if it gets out of hand, genetic engineering will merely join a whole catalog of massive abuses of the earth's natural environment. Will we change now? Don't count on it.

Some observers are already asking plaintively, too, "Why can't we leave Mother Nature alone?"

The answer, of course, is that man has never left nature alone, and never will. Man, himself, is a part of nature and, as such, it is normal that he exploits what he finds, where he finds it. This shouldn't suggest, however, that exploitation of nature's riches needs to be unwise. As products of nature, men and women are still perilously subject to all the natural laws. Even at the dawn of genetic engineering, humans must still obey all of the immutable laws of physics and chemistry. Genetic engineering is basically merely a means of exploiting these physical laws, using them to steer living things in new directions. It amounts to taking the tools—the genes—that Mother Nature has provided and putting them into productive new uses. Sometimes they're going to cause problems, or perhaps not work at all, but it's clear that humans are now beginning to play with real genetic fire—or you might call it a genetic version of Russian roulette—because dangers do exist.

It's also being asked—or actually stated by environmentalists like Barry Commoner—that "Mother Nature knows best."

Well, perhaps, but the birth of a mongoloid child doesn't seem, in the eyes of some scientists, as though nature's doing the best job. Looking at ruined fields and starving people, agricultural scientists answer that disease and famine are also a large part of nature, a part that people, animals and crops could well do without. Brain specialists, watching the slow mental decline of an otherwise healthy person, would answer, "No, indeed, nature doesn't always know best."

And the goals of genetic engineering? They're many and varied and they seem sometimes to change with the wind. Depending on who's setting goals, genetic specialists can foresee the day when hunger, for all practical purposes, is finally abolished. They also look ahead to the time when disease disappears, and to the time when inborn genetic diseases are cured with a dose of a friendly microbe designed to deliver a few new genes to a body made of faulty cells. They can look forward, perhaps, even to the time when mental institutions can close their doors forever.

Heady stuff, but not at all forbidden.

Industrial scientists, too, have their goals, looking ahead to better, purer chemicals produced cheaply in clean factories, goods actually manufactured by genetically engineered microbes. Wastes will be used as raw materials. Sewage will provide energy. Waste paper will be turned into sugar. Medicines, now scarce and expensive, will be cheap and abundant. New foods, produced by microbes devouring waste materials, may be more abundant and more nutritious than many natural foods. Pesticides, produced and tailor-made to order by fungi and bacteria, will keep insects at bay without poisoning the world we live in.

These, then, would be a few of the broad goals. Not all will be realized soon, nor all at once, and some may never be reached. But they do represent reasonable estimates of what can be—and what is being—achieved.

Freedom? That will probably depend on how one defines it. The Mainland Chinese, free now from rampant disease and hunger for the first time in history, would loudly proclaim and with good reason, their own freedom. Americans, blessed with a level of affluence and abundance unheard of in history, look instead at the Chinese experience as tantamount to bondage, inflexible near-imprisonment.

For the people of Africa, Latin America, and the rest of Asia, who spend their days scratching desperately for any food they can grow or find, the budding science of genetic engineering offers eventually—and perhaps for the first time—a full stomach and a healthy body.

But not yet. Even the promised miracles of genetic engineering —like the proposed miracles of that Green Revolution—aren't going to overcome the awesome effects of a rampant birthrate, of overcrowded slums, of poor sanitation and lack of hospitals. Until governments care and act, and until people everywhere do likewise, the problems of the world's desperately poor will persist.

On a worldwide basis, food is abundant. But in the world market it's sold to the highest bidder, not necessarily to the people who need it most. Millions of tons of U.S. grain, for example, are sent yearly to Japan, to Europe, and to Russia, where diets are already relatively rich. Only some 2 percent of United States surplus grain ever gets to the people who really need it, to those who don't have the money to buy it. So even now, despite episodes of starvation in Ethiopia, in Bangladesh and elsewhere, there is no real problem of food production. It's a problem of food distribution and storage, and it's a problem of sadly misplaced values.

How misplaced, indeed, should be evident when Americans, if they spent only half as much on the poor in other nations as they spend feeding squads of dogs and cats, could probably solve that problem overnight. Most American dogs living with middle-class families, indeed, thrive on a better, more protein-rich diet than most of the poor who live in the hunger zones of the world. This is a disgrace which, perhaps far in the future, is going to be difficult to live down.

What one would wish, as genetic engineering begins coming into bloom, is that some orderly way will be found, based on careful reasoning, for deciding what should be done first. Unfortunately, in Western nations, at least, no such mechanism yet exists outside of industry where some of the decisions will be made solely in the marketplace. Some sort of guidance in the United States, and probably in England, will also be coming from funding agencies as they allocate money for research. The National Science Foundation and National Institutes of Health in America—and the Medical Research Council in England—will be deciding to push research in special directions.

It's almost certain, too, that in some instances the promise—and more likely, the peril—of genetic tinkering will be seized and squeezed painfully as a viable political issue. Hopefully it won't amount to anything as shrill and damaging as the antics of the McCarthy era, but the opportunity does exist to scare the wits out of people with horror stories about what might happen. This has already started, for example, in Cambridge, Massachusetts, where a politically awake city council has won a brief moratorium on some recombinant DNA work at Harvard University and the Massachusetts Institute of Technology. If the Cambridge debate goes far enough, and if the arguments become shrill enough, it may provide the opportunity for some of Boston's opportunistic politicians to ride the issue to higher office. One of the surest ways to gain attention in both Cambridge and Boston is to bait the universities, lamenting loudly in public about the lost tax base and the irresponsibility of academe.

At higher levels of the political structure, too, the debate will probably be taken up, especially as the powers of genetic tinkering become more widely known and as the impact of such techniques come closer to being felt. It will probably happen that something akin to Nixon's war on cancer will be ordered in Washington, and it will probably serve mostly to unbalance the whole research effort. Worse, a crash program involving genetic engineering—such as growing up the first cloned man—might provide more room for disaster than for success.

The first results of true genetic engineering, instead, will obviously involve the microbes—as with General Electric's oil-eating *Pseudomonas* bacterium—and most of the work in the next few years will still involve unscrambling many of the scientific details rather than producing spectacular new creatures. After the science and technology have been well worked out, however, some interesting choices about what to do will be possible. Industry will be looking toward the profit picture—that so-called "bottom line"—so not much help can be expected from this quarter toward production of new foods and products from plants growing in the tropics, where help is needed most. As a result, outside help will be required, and it will probably again be funneled through those two international research centers, one for wheat in Mexico and one for rice in the Philippines. The first experiments at these two centers will probably be aimed at overcoming some of the deficiencies of their present Green Revolution plants, making them even more useful, and more acceptable, in the areas of Asia and Latin America where they are meant to thrive. What goes into these vigorous crop plants, of course, will depend on what has become available through laboratory research. If such research includes efforts to transfer the gene or genes for nitrogen fixation from the legumes into other plants, and if it finally does succeed, then it's probable this work will quickly be taken up in these two plant research centers, since a shortage of expensive nitrogen-rich fertilizer is what seriously crippled the Green Revolution.

With all this in mind, then, one should ask who in this world is going to be benefiting from genetic manipulation, and who is apt to be hurt?

The benefits may be relative. Scientists will benefit through advancing their own careers, and they can be justifiably proud if, through tinkering with the genetic codebook, they can help solve the world's worst problems of hunger and disease. Obviously, too, the people who receive extra food and thus avoid starvation will be beneficiaries of genetic tinkering. But such benefits will only persist if population growth in Asia, Africa, and South America is somehow finally controlled. If population growth remains unchecked, then any benefits from genetic engineering will soon evaporate.

It can be expected, too, that American industries, especially those involved in agriculture and production of chemicals, will benefit enormously; probably because they'll be holding the patents on new processes, new bugs, and new varieties of food plants that are abundantly productive. In this case, of course, it's the rich once again getting richer.

But then, too, there should also be some disruption of industry. If, for instance, that famous nitrogen-fixation gene is ever transferred into corn and other normally fertilizer-hungry crops, there will probably be a significant decrease in demand for fertilizers. And, with the increases in food production that can be expected, farmers may end up receiving less money even though they're producing more food. Thus price controls may still be with us.

Another disappointing prospect is that some of this awesome power in American science will be funneled into trivial areas. Look, for example, for advertisements someday for new types of grass, new lawns, which require no fertilizer. There will be Bermuda grass, for instance, incorporating the nitrogen-fixation gene. Unfortunately, the American consumer has the dollar power to override some of the more obvious, necessary uses of technology, drawing the energy, talent and time of sophisticated industries into building silly gadgetry for automobiles, spray cans, and toilets and telephones designed to look like they're going 100 miles an hour. This is a bit appalling when the need for these is compared to the real problems that must go begging, unresolved. Again, it's that nasty problem of priorities.

Who gets hurt? Unfortunately, the chance exists that everyone can get hurt if genetic engineering goes astray. Any "worst-case analysis" can dream up horror-filled scenarios about hordes of new man-made microbes killing millions of people. Such possibilities do exist, but it would seem much more likely that there will be series of small accidents, and the damage will be of much smaller scale. More likely, too, there will be a series of disappointments rather than disasters. Scientists, perhaps, will design and introduce some new supercrops in famine areas, only to see them fall flat. It's almost predictable that something coming right out of the natural environment will arrive to cause some disappointing setbacks, some real defeats. Scientists will not be able to forget that Mother Nature still has her evolutionary powers intact, and that she will constantly be throwing up new challenges to the best that

can be built. This already occurs periodically when a new mutation of an old influenza virus comes along and finds a population
unprotected. This will continue to happen, although scientists'
response time will probably grow shorter.

The problem of too much genetic uniformity will remain too.
Uniform crops, uniform animals, and even uniform people, eventually will be offering a much wider, easier-to-hit target for some
bacteria, viruses and fungi which might be able to find an important chink in someone's immunological armor. Once a disease
organism is turned loose in a susceptible and very uniform population, its spread is faster and farther than normal. The best recent
example of this was the wave of corn blight that rushed across
much of the United States corn belt one summer, wiping out about
15 percent of that valuable crop.

All this hemming and hawing back and forth over risks and
benefits would suggest, then, that the potential for important,
history-making advances can be seen in genetic engineering,
which is expected to supply the world with productive new types
of plants, animals, and even people. Yet along with such benefits
will come chances that problems will arise, that the mistakes made
will be bigger and more hard-hitting, possibly, than anything
we've encountered before. One can indeed be sure that some mistakes will be made. We are still human.

Do all these potential benefits outweigh the possible hazards?
It's unfortunate, but for now the prospects on both sides of the
argument are still so uncertain that a firm answer is impossible.
The risks, up to a point, are mostly hypothetical, based on a series
of sometimes wild "what if" questions. Many of the biologists and
other specialists involved now in the recombination of DNA molecules do agree that some risk exists, but they all point out that no
one is yet able to determine *how much* risk there is. The alter-
natives could range from creation of a new, deadly, unstoppable
plague that would sweep worldwide; or, scientists may just create
a bunch of laboratory-built organisms which are so benign, so
weak and imperfect, that they can't compete in the wild with
other creatures which have had billions of years of evolutionary
experience in which to sharpen up their competitive skills. Still,

there is enough nervousness about the possible hazards so that researchers agree there are a few experiments which should clearly not be done—at least pending some answers.

As for the benefits expected from genetic manipulation, these are equally difficult to assess because of the many uncertainties that remain. It seems almost certain, however, that microbes, given new talents through insertion of new genes, will provide some important advances for industry. In agriculture, however, there is nothing—beyond hope—to assure that the genes responsible for nitrogen fixation will ever be successfully transferred into wheat, corn, and rice. But if they are, the benefits would be so enormous and so important that at this point not to try, out of fear or timidity, would be almost a crime. The same is true in medicine. The chances for alleviating human suffering through new medical and genetic techniques are so great—but certainly not assured—that not to try, not to push research in this direction, would be immoral, some people would argue. Indeed, we are already beginning to catch up on some of the genetic diseases through screening, amniocentesis, and abortion, and not to follow up in this direction toward the cure of genetic diseases would probably be wrong. Still, it would also be wrong to pour enormous resources into curing a few people of their genetic disabilities—however pitiful—at a time when most people can't even get treatment for the diseases we already know how to cure. What's needed, then, is to strike a reasonable balance, in terms of allocation of resources, between advanced research and the efficient delivery of health care.

What, then, should be done about genetic engineering? Should citizens in the United States, Britain, Sweden, France, Japan, and Russia mobilize to halt this work, stop it before some deadly new microbe escapes? Or should citizens ignore the possibilities and trust in scientists' good motives and strong assurances that everything will be all right?

The answer to both questions, of course, is no. Citizens, worldwide, will have to find ways to keep informed about what's going on in the laboratory, and they must find ways to prohibit the performing of foolish and dangerous experiments. Determining which experiments are foolish, dangerous or both—separating these from the experiments which are useful and wise—however, is something the public can't be expected to do. Nor can this be done by the public's fallible representatives, the politicians. This says, then, that the scientists will have to be involved in regulation of such research, but—as pointed out by that Cambridge, Massachusetts, group, Science for the People—scientists alone should not be left in control. Such overspecialized groups often too easily lose sight of the bigger picture, including the interests and the survival of society. Also, the important decisions regarding what is to be done, and how and where it is to be done, must be made in the public arena, open to everyone with anyone free to object. This should not mean, however, that every time a biologist wants to run an experiment he should be asked to seek permission from Congress. Workable mechanisms must be set up to oversee genetic research without restricting it unnecessarily, without tying research up in red tape.

Further, on the serious question of safety, it now seems clear that some experiments—such as those requiring P-3 or P-4 containment—should be done someplace else, away from cities, away from crowded university campuses. There is no sense in unnecessarily endangering the population of a major metropolitan area by doing possibly dangerous experiments right in the city. The biologists hoping to perform these experiments will complain that it's inconvenient to travel far away just to run their experiments. But inconvenience and discomfort cannot begin to equal the "inconvenience" a city's people would experience if there were a significant accident. And, as some scientists expect, there are almost certainly going to be some accidents. Some of these invisible, strange new bugs are going to get out sometime. Whether they're actually lethal makes little difference in this argument. They can be sublethal—just sickening—and still not be worth the risk.

Perhaps the suggestion proposed by Dr. Robert Sinsheimer, chairman of the department of biology at the California Institute of Technology, is wisest. It has the ring of truth, and of experience, even though some of Sinsheimer's fellow biologists are ready to dispute his logic. Sinsheimer says that the most dangerous or potentially dangerous research should be deported from the cities. These experiments should be taken far into the countryside, and they should be performed there inside some brand-new ultra-secure laboratories.

An immediate example that comes to mind is the U.S. Army's old biological warfare research laboratory at Fort Detrick, Maryland. This facility is already prepared to handle dangerous organisms with some safety—but not with perfect safety—and it might easily be prepared, modernized, to handle recombinant DNA work. Also, since numerous new laboratories are going to be built for this work, probably by various universities with government support, it would make sense to build them in remote areas, away from population centers. The cost of construction should not be much more, and the costs of transportation and housing for visiting scientists would not be enormous, and probably worth it in the long run. Biologists aren't used to having to do this, but astronomers have been doing it for years, traveling to remote locations to use big telescopes, then bringing their data home again for analysis. In addition, compared to the costs of a serious epidemic breaking out in Boston or Berkeley, the expense of transporting biologists off to distant laboratories would be minuscule. The whole point, of course, is that if the research is worth doing —and it seems to be worth doing—then it ought to be done where it's safest, where the dangers can be minimized.

Additional research should be done as soon as possible to assess what the dangers really are, and the experimental work should be halted if these results show the risks seriously outweigh the benefits.

There are some even weightier questions about genetic engineering which deal with the social impact of this new science. When families are given the choice someday, for instance, to decide the sex of their offspring in advance, what will be the result? Some surveys report that many families prefer having a boy first, a girl second. There are certainly some well-defined psychological consequences to being the first or second child in a family, and it is known that these consequences act very differently according to the sex of the child. Thus it's interesting to speculate on what the results might be if most families did choose to have a boy first. Would we end up with a society in which no women had been first children; in which no men were last? And what would be the consequences now if all women had grown up in the shadow of older brothers; if they all grew up with the realization that they had been second choice?

Such questions are obviously unanswerable now, but the implication is that such things may someday become important in unseen, unexpected ways.

And then there's the prospect of cloning. What happens to that first cloned individual? This first cloned person would truly be the new Adam or the new Eve, an orphan in a new and different sense. How would such a person fit into society? Would he or she soon be interacting with a whole herd of identical twins, his or her fellow clones? And the family? Would it disappear, a lost institution?

Sinsheimer, who has been speaking out more and more forcefully on these issues, asking some of the deepest, most far-reaching questions, wrote in Caltech's alumni magazine, *Engineering and Science,* asking:

> Do we face an ambush—or an epic opportunity? A forbidden universe—or the long-sought land, the goal toward which evolution has been striving for 5 billion years . . . ?
>
> We have, over the past few centuries, achieved a very considerable mastery over our physical world, and many are less than pleased with the results. We can now foresee—through our new insight into the bases of life—a growing mastery over our biological world—and that includes us—and many are terrified at the prospect.
>
> They are not without reason. Much of the despair of our times stems from the realization that at last, after all the toil and all the invention, all the savagery and all the genius, the enemy is "us." Our deepest problems are now "made by man."

Sinsheimer also warned:

> . . . we cannot escape the need to make crucial value judgments. Man *is* a social animal, the product of his culture as well as his genes. . . .

(FLOYD CLARK, CALTECH)

Dr. Robert Sinsheimer, chairman of the biology division at the California Institute of Technology, has become one of the most outspoken advocates of caution in genetic manipulation experiments. Sinsheimer, who was one of the first prominent biologists to begin thinking and speaking about a future that includes genetic engineering, has concluded that potentially dangerous types of experiments should only be done in remote, ultrasafe laboratories, well away from population centers.

For what kind of society shall we select genetic characteristics? To what kind of society might the simple act of gene selection lead? Are some matters best forever left to the winds of chance? Or is that a failure of nerve and a denial of the human experience? Is it somehow inhuman to design a man?

Once again, then, there are no answers—or worse, there are too many small, half-answers, too many partial opinions, too much conjecture; mountains of prejudice and whole valleys of superstition. It may be unfortunate, but it's already evident most of these important questions are not going to be answered in advance.

Index